Modern optical techniques have a huge range of potential applications in signal processing and in the interconnection of digital computing systems. This book provides a detailed review of the key issues which must be addressed in the design, evaluation, and implementation of practical systems for processing and optical interconnection.

Although the advantages of optical processing have been shown in the laboratory, it is still necessary to determine whether or not they can be retained in full-scale applications, and each chapter of the book is devoted to a different aspect of this field. Considerations such as the computer modeling of optical design limitations, the size and noise characteristics of optical modulators, and the relative merits of free-space and guided-wave optical technology in different processing systems, are all discussed in detail.

In examining the important limiting factors and characteristics of practical optical processing systems, the book will be of great interest to optical researchers and designers, and to anyone wishing to learn about the basic techniques of optical processing.

CAMBRIDGE STUDIES IN MODERN OPTICS

Series Editors
P. L. KNIGHT
Department of Physics, Imperial College of Science, Technology and Medicine

A. MILLER
Department of Physics and Astronomy, University of St Andrews

DESIGN ISSUES IN OPTICAL PROCESSING

DESIGN ISSUES IN OPTICAL PROCESSING

Edited by

JOHN N. LEE
US Naval Research Laboratory

Published by the Press Syndicate of the University of Cambridge
The Pitt Building, Trumpington Street, Cambridge CB2 1RP
40 West 20th Street, New York, NY 10011-4211, USA
10 Stamford Road, Oakleigh, Melbourne 3166, Australia

First published 1995

Printed in Great Britain at the University Press, Cambridge

A catalogue record for this book is available from the British Library

Library of Congress cataloguing in publication data

Design issues in optical processing / edited by John N. Lee.
p.   cm. -- (Cambridge studies in modern optics ; 16)
Includes index.
ISBN 0-521-43048-8
1. Optical data processing.   I. Lee, John N., 1944-
II. Series.
TA1630.D47   1995
621.39'1--dc20   94-19147   CIP

ISBN 0 521 43048 8   hardback

KT

Contents

List of contributors

Chapter 1
How optical computers, architectures and algorithms impact system design
DAVID CASASENT
Department of Electrical and Computer Engineering, Carnegie Mellon University, Pittsburgh, PA 15213, USA

Chapter 2
Noise issues in optical linear algebra processing design
STEPHEN G. BATSELL
Naval Research Laboratory, Washington, DC 20375-5337, USA
JOHN F. WALKUP and THOMAS F. KRILE
Department of Electrical Engineering, Texas Tech University, Lubbock, TX 79409-3102, USA

Chapter 3
Effects of diffraction, scatter, and design on the performance of optical information processors
R. B. BROWN and JOHN N. LEE
Naval Research Laboratory, Washington, DC 20375-5338, USA

Chapter 4
Comparison between holographic and guided wave interconnects for VLSI multiprocessor systems
MICHAEL R. FELDMAN, JEAN L. CAMP, ROHINI SHARMA and JAMES E. MORRIS
Department of Electrical Engineering, University of North Carolina, Charlotte, NC 28223, USA

Chapter 5
High-speed compact optical correlator design and implementation
RICHARD M. TURNER, KRISTINA M. JOHNSON
Optoelectronic Computing Systems Center, Boulder, CO 80309-0525, USA

and
STEVE SERATI
Boulder Nonlinear Systems Inc., 1898 S. Flatiron Court, Boulder, CO 80301, USA

Chapter 6
Optical and mechanical issues in free-space digital optical logic systems
F. B. McCORMICK
AT&T Bell Laboratories, 263 Shuman Boulevard, Naperville, IL 60566-7050, USA
and
F. A. P. TOOLEY
Department of Physics, Heriot-Watt University, Riccarton, Edinburgh, EH14 4AS, UK

Preface

The design of optical hardware for signal and image processors and in computing systems can be expected to become an increasingly important activity as optics moves from the laboratory to actual usage. Over the past decades a variety of concepts have been developed to exploit the capabilities of optics for ultrahigh-bandwidth and massively-parallel processing such as for analog front-end operations, image-processing and neural-net applications, high-bandwidth interconnection of electronic systems such as computers and microwave subsystems, execution of logic and other nonlinear operations for routing of optical interconnects, as well as for general reduction of hardware size, weight and power consumption. Original concept investigations usually identify theoretical advantages of performing various tasks with optics, but one then needs to address whether the advantages can be maintained as one addresses various new issues involving usage in a real-world environment; e.g., can demonstrations of optical processing speed and parallelism, optical routing, and communication bandwidth be retained in full-scale application? Addressing these questions very often requires an optical approach that employs different algorithmic/architectural approaches than conventional alternatives; hence, new architectures are devised to utilize maximally the strengths of optics. Finally, the capabilities of optical technology may dictate specific optical hardware designs.

It is therefore difficult to predict whether a specific optical hardware configuration will still possess the originally identified advantages. Although full-scale demonstrations can provide a direct answer, careful design and analyses must be performed to ensure an effective demonstration. The design and analysis stage can involve fundamental issues that are quite removed from the original optical-processing concept.

It is often found that multidisciplinary approaches are required in practice. For example, it is often required to design unique optical systems that preserve dynamic range, bandwidth, resolution, etc., but that also must be very compact, and tolerant of temperature variations, shock, and vibration. As a possible consequence, even if basic optical components and devices such as spatial light modulators have been demonstrated, the optical design may require parameters, e.g., dimensions and tolerances, that are quite different than those available or those that are feasible. Hence, further device development is usually required beyond the technology-feasibility stage, involving careful materials and fabrication engineering. Finally, the optical processor design itself may have to be modified iteratively to be made compatible with thermo-mechanical requirements. In parallel with design and construction for demonstration, other considerations need to be evaluated. Such considerations include whether scale-up of parallelism degrades performance, the physics of the noise contributions from various optical devices, limits on numbers of parallel channels due to crosstalk mechanisms, etc. Any optimization must intimately involve details of the specific proposed use. Hence, the different chapters examine various issues in optical processing applications.

This book covers a range of the important areas in optical processing and optics in computing, emphasizing algorithmic and design issues affecting hardware implementation. The chapter organization is such that each covers a distinct area, as described in the following summary list; but there are important interrelationships between chapters, e.g., the design tools described in one chapter can be of value to topics covered in another.

(1) Optical architectures and algorithms for processing (Chapter 1 by Casasent). Optical techniques have been shown viable for performing one- and two-dimensional (1-D and 2-D) linear integral transformations, e.g., Fourier transforms and correlation. Such operational primitives might be useful in larger computational algorithms.

(2) Acousto-optical signal processing (Chapter 1, Casasent, and Chapter 3 by Brown and Lee). 1-D acousto-optic devices (Bragg cell) have been shown to be effective means for input of electrical information onto light beams for 1-D, multichannel 1-D and some 2-D linear optical processors, and have been the basis of many high-performance laboratory demonstrations.

(3) Optical image processing and pattern recognition systems (Chapter

5 by Turner, Johnson, and Serati). A classic example is the use of the Fourier-transforming properties of a lens and a 2-D spatial light modulator to perform scene correlation.

(4) 1-D and 2-D spatial light modulators device characteristics (Chapter 2 by Batsell, Walkup, and Krile).

(5) Optical interconnects in digital computers (Chapter 4 by Feldman, Camp, Sharma and Morris). Architectures and hardware are being considered for optical interconnections between nodes of computer networks, computer subsystems such as microprocessor boards, and even electronic chips, with the first closest to reality.

(6) Optical digital logic (Chapter 6 by McCormick and Tooley). Non-linear operations such as logic operations are easily performed with electronic devices, but if they can be performed directly in the optical domain there is the potential to alleviate bottlenecks and signal-to-noise degradation due to optical to electronic conversions, and to allow massively-parallel 2-D computation, especially for new architectures such as for optical switching and neural networks.

The following is a brief synopsis of the major issues addressed by each chapter.

Chapter 1 by David Casasent shows how clever algorithmic and architectural techniques can take the basic Fourier-transforming and correlation capabilities of optics and produce a programmable general-purpose multifunctional optical image processor. The significance of this is that very similar hardware is used to perform an extensive repertoire of image-processing operations, and a full reoptimization of components and design need not be performed for every specific function within this repertoire. There is a common set of hardware: laser diode arrays, acousto-optic cells, and perhaps of most significance computer-generated holograms which may either by fixed or vary with time using a 2-D spatial light modulator. The architectures, or arrangements of the hardware, can be one of or a combination of two basic classes – space-integrating or time-integrating. The former can be achieved by techniques such as focusing of an optical beam within an aperture, the latter can be achieved by techniques such as accumulation of charge from photodetectors over a specific time interval. Among the repertoire of operations that are needed in a fully functional image-processor system are: transforms for feature extraction such as the Fourier transform and the Hough transform, and linear-algebraic filters for distortion-invariant performance; low-level image-enhancement operations such as edge-enhancement and image filtering such as morphological filtering; additional operations such as wavelet and Gabor that may be needed along with the preceding for effective detection; and finally, certain classes of neural net classifiers are needed.

Chapter 2 by Stephen Batsell, John Walkup and Thomas Krile addresses noise issues in optical linear-algebra processors. Such processors have played an important role in the history of optical processing. Linear algebra is a means of describing a large variety of mathematical operations, including those for signal processing and image processing as described in Chapter 1. For example, the kernel of many integral transforms, including the Fourier transform, can be described as a vector–matrix product. The optical linear-algebra processors considered here use analog intensity to perform numerical calculations, but are not considered general numerical computers. Analog optical signal processing can serve an extremely important niche role in providing ultrahigh-performance coprocessors for systems. Both fixed and signal-dependent noise sources are analytically examined. The effect of hardware performance on these noise sources is obtained from analytic expressions for signal-to-noise and accuracy that include noise contributions from lasers and other light sources, light modulators, and photodetectors. Experimental results are then described, especially for available spatial light modulators. As an example of the effect of device construction, it was found that the built-in temporal modulation of liquid-crystal spatial light modulators greatly reduces the signal-to-noise ratios of the products.

Chapter 3 by R. Bernard Brown and John N. Lee describes computer-aided design (CAD) techniques, simulation approaches, and CAD tools for optical processors, along with specific examples. Processors need to be designed with a high dynamic range, a large number of parallel channels (i.e., large space-bandwidth), and high optical-phase and amplitude precisions. Further they must be designed to be tolerant of thermal variations and vibrations. Achievement of all these goals requires careful consideration of a number of effects – e.g., optical diffraction, optical scatter, stray light, and optical aberrations – in the design of Fourier-transform and relay-imaging systems. This chapter provides design examples for acousto-optic microwave radio frequency (rf) spectrum analyzers (or channelizers) and for space-integrating acousto-optic temporal signal correlators. The former examines effects that lead to degradation of optical channel bandshape, and the latter example focuses on simulation of contributions to errors in the optical correlator output, especially due to Fourier relay imaging systems. A number of commercial design tools are described that have been adapted to address optical-processor requirements. Although specific commerical packages, such as Code V^{TM}, are mentioned extensively, it should be emphasized that it is the general approaches used by such design packages that are of interest; this is especially important to realize since newer, improved packages are expected to become available. The design packages as they are applied to optical processors must adequately handle important factors such as diffraction, small-angle optical scatter, and stray light over a large dynamic range, and several such packages may have to be employed in an integrated fashion.

Chapter 4 by Michael Feldman, Jean Camp, Rohini Sharma, and James Morris considers how to utilize optics for interconnecting very large arrays of digital microprocessors. Such multiprocessor architectures are widely considered as necessary to push digital-computation capabilities beyond the limits of conventional serial computers. However, a major roadblock is the high-bandwidth data interconnections needed to implement such architectures effectively. This chapter considers three popular interconnection schemes: the mesh, the hypercube, and the fully-connected scheme – the last being particularly difficult to implement electrically for large arrays. Comparison is made between guided-wave interconnects and free-space interconnects via holograms in terms of connection density, interconnect latency and power dissipation. In general, as the interconnectivity becomes higher (e.g., hypercube or fully-connected), and the number of processors in the array increases, the free-space optical interconnects are seen to gain advantage over the guided-wave approaches.

Chapter 5 by Richard Turner, Steve Serati and Kristina M. Johnson describes a high-speed compact optical correlator design and implementation using a 2-D spatial light modulator (SLM) for scene correlation. Although scene correlation is among the first uses of optical processing, it is only recently that fully-engineered demonstrations have been achieved. Among the important design issues are the influence on optical-system correlation performance of SLM device size, pixel size, device contrast ratio, device flatness and uniformity, pixel fill factor, and framing speed. Although the particular implementation uses the ferroelectric liquid crystal and silicon SLM, many of the results can be extrapolated to any type of SLM.

Chapter 6 by Rick McCormick and Frank Tooley covers an emerging area of research: to exploit the inherent massive parallel-channel capability of 2-D optics by interconnecting very large arrays of simple logic elements at very high bandwidth. The basic device considered has a bistable optical transfer function and plays the same role as a transistor in conventional electronics. The device requirement can be satisfied by either optically nonlinear media within a Fabry–Perot etalon, or hybrid devices such as the self-electro-optic device (SEED), or light-emitting devices such as laser diodes or pnpn optical thyristors. The choice of device is dictated by consideration of system-related issues, such as the size of the detector and modulator/emitter elements, element pitch, and device switching energy, power and speed. For example, nonlinear etalons are unpixellated so device size and pitch are freely chosen, but they generally require more switching energy than SEED devices. The system design example in this chapter is drawn from work with symmetric or S-SEED devices. Although much research has gone into researching the fundamental feasibility of performing optical logic with devices such as S-SEED elements, this chapter makes it clear that unless extreme care is taken in the optical design, there can be severe limits on the practical size of the logic arrays that can be handled because of the need for multiple imaging operations. Reduced optical system complexity, however, requires increased levels of sophistication at each SEED.

J. N. Lee

US Naval Research Laboratory

1

How optical computers, architectures, and algorithms impact system design

DAVID CASASENT

Carnegie Mellon University, Department of Electrical and Computer Engineering

1.1 Introduction

When a scientist or engineer is provided with a given processing problem and asked whether optical processing techniques can solve it, how does he answer the question and approach the problem? To address this, he first considers the basic operations possible. They are often not a direct match to a new problem. Hence new algorithms arise to allow new operations to be performed (often on a new optical architecture). The final system is thus the result of an interaction between optical components, operations, architectures, and algorithms. This chapter describes these issues for the case of image processing. By limiting attention to this general application area, specific examples can be provided and the role for optical processing in most levels of computer vision can be presented. (Other chapters address other specific optical processing applications.)

Section 1.2 presents some general and personal philosophical remarks. These provide guidelines to be used when faced with a given data processing problem and to determine if optical processing has a role in all or part of a viable solution. Section 1.3 describes optical feature extractors. These are the simplest optical systems. This provides the reader with an introduction and summary of some of the many different operations possible on optical systems and how they are of use in image processing. A major application for these simple systems, product inspection, is described to allow a comparison of these optical systems and electronic systems to be made.

Section 1.4 addresses the optical correlator architecture, since it is one of the most powerful and most researched architectures. The key issues in such a system used for pattern recognition are noted. These include: how do you achieve distortion-invariant recognition, which spatial light modulator (SLM) do you use and how do you alter the architecture to allow use of available devices, and what do you do if the SLM is binary and the data is analog? Optical correlator hardware has significantly matured and thus this architecture merits considerable attention.

However, if a correlator is to see use in a full image processing system, many operations beyond distortion-invariant object recognition are necessary. Section 1.5 notes these other required operations (the levels of computer vision) and how new algorithms allow each to be addressed. These operations are all possible on the same basic architecture, an optical correlator. Thus, a multifunctional general purpose optical system results. Such general purpose optical processors are felt to be essential if optical processors are to see practical use.

I conclude (Section 1.6) with brief remarks on optical neural nets and their role in image processing. A summary and conclusions are then advanced in Section 1.7.

1.2 Philosophy

When faced with a given processing problem and the question of whether optics has a role in its solution, one should ask at least three basic questions.

You should look at the operations that optics can perform. It is not enough that it can do the operations needed for the problem (Fore, 1991). Optics should offer a speed advantage over electronic approaches. The advantage would probably have to be at least a factor of ten before one could sell it. Cost is another major factor, but this is beyond the scope of these brief pages. Size, weight, and power consumption are other factors in some (but not all) applications. Several studies (Fore, 1991) have shown that optical correlators offer size, volume, speed, and power dissipation advantages over comparable electronic systems. Thus, our focus on them seems valid. In the area of image-processing optical components, architectures and algorithms are sufficiently developed, and thus this area is considered. For optical logic and numeric processors, these answers are not yet clear.

One must look at the application and consider *how the operations can be performed optically*. One generally develops an algorithm that does not require operations that an optical system cannot easily achieve. Specifically, an algorithm that requires many conversions between optical and electronic processing is not attractive. However,

one need not become obsessed with performing all operations optically.

An arrangement of optical components is thus made that can perform the necessary operations. However, *the devices postulated must
exist*. If they don't, they must be easily developed. New optical
architectures and systems generally evolve by modifying devices developed for other purposes: displays, disks, fiber optic communications,
printers, etc. Optical processing is generally not yet able to support the
design of elaborate devices specifically needed only for it.

Other remarks that I personally advocate are now advanced. One
should not copy an electronic processor or algorithm, since its design
was inevitably based on electronic not optical components and functions. I feel that optical processors should be analog, since therein lies
the power and uniqueness of optics. The accuracy possible on analog
optical systems must be sufficient for the problem, or the algorithm
must be chosen to be robust to analog accuracy components. Images
are analog data; and image processing, pattern recognition, and neural
net applications and algorithms generally satisfy these requirements.

Optical processing researchers must thus, by definition, be inno-
vative in the algorithms and architectures they develop. As we shall
see, this has been very true.

1.3 Optical feature extractors

Nearly everyone is familiar with the fact that a coherent optical system
can form the 2-D (two-dimensional) Fourier transform (FT) of input
data on an optical transparency. This section briefly notes a number of
other useful descriptions of an input object that can be optically
produced. This is intended to show the versatility of optical systems,
the growing repertoire of operations they can perform, and how they
are of use in image processing.

Figure 1.1 depicts the general problem and approach. A single
object is input. Different features describing it are calculated and these
are fed to a classifier. The output can be answers and information such
as: is the input a good or bad product; if it is bad, what type of defect
is it; what is the identity of the object; what is its orientation; etc.? The
major computational load is calculation of the features, thus this is a
viable portion of the system for which to consider the use of optics.
Feature calculation requires global operations with many operations
per input pixel, analog accuracy suffices for most cases, and the same

Figure 1.1 Block diagram of an optical product inspection system.

set of features can be useful in a wide range of different problems. Thus, this operation seems appropriate for a rigid non-flexible optical architecture. The classifier must be different for each application and thus classifiers are more suitable for electronic implementation. Thus, a viable hybrid optical/digital system seems realistic. In product inspection, lighting and illumination are controlled and thus no preprocessing is generally required and hence a simple system should suffice.

Figure 1.2 shows the well-known optical FT system with a wedge-ring detector (WRD) placed in the output plane. This WRD device consists of wedge and annular (half ring) shaped detector elements (Lendaris & Stanley, 1979), or the equivalent may be synthesized with a spinning disk with properly designed apertures (Clark & Casasent, 1988). A key point not generally acknowledged in the optics community is that digital hardware can also perform the 2-D FT at TV frame rates (30 objects/s). However, the cost of the digital system exceeds that of the optical system, if an inexpensive input SLM exists. This is now the case with liquid crystal TV devices (Liu, Davis & Lilly, 1985) as the SLM (if they are suitably modified to be optically flat and to have no residual images; i.e. active erasure, not decay, is needed). Producing an FT alone is not sufficient. Specifically, there are as many FT samples as there are input samples, and it is not easy to analyze all of them in real time. Thus, the WRD provides another vital property, data compression. With 32 wedge and 32 ring detectors, only 64 outputs need be analyzed for each input object. These features have other well-known properties that are vital for product inspection: the

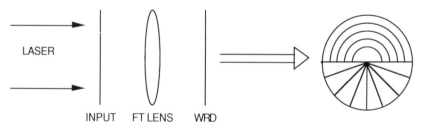

Figure 1.2 Optical WRD system (Casasent, 1991).

feature space is shift-invariant (since the intensity of the FT is detected, the wedge features are scale-invariant and their pattern shifts with image rotations, and the ring outputs are rotation-invariant and their pattern shifts with image scales).

A simple lens forms the optical FT. To produce other optically generated feature spaces, computer generated hologram (CGH) elements (Lee, 1992) can be used. With them, the resultant optical system remains simple (and hence inexpensive). Such architectures thus resemble Figure 1.3. If the CGH has a transmittance pattern that is a set of N cylindrical lenses each at a different rotation, then the output pattern is the Hough transform (HT) of the input (Richards, Vermeulen, Barnard & Casasent, 1988). This feature space consists of peaks of light. The location of each peak denotes the orientation and position of each line in the input and the height (amplitude) of each peak denotes the length of each line. Extensions to curves, etc. are also possible with generalized HTs. This example is included because it demonstrates how CGHs (using integrated circuit lithographic techniques) allow new optical functions to be implemented very cost effectively. Another very vital and generally not appreciated property of an optical system is that it automatically performs interpolation. This has been used in the optical HT system and comparison of the calculated HT with digitally calculated results shows that *the optical system is actually more accurate* due to digital sampling problems (Richards & Casasent, 1989).

The WRD FT and HT feature spaces have proven very useful in product inspection and object identification. Other optically generated feature spaces include chord distributions, arc angle/length features of a contour, Mellin transforms, polar-log transforms, and moments (Casasent, 1985).

An optical system that computes the moments of an input object is shown in Figure 1.4. It is not competitive with digital chips that

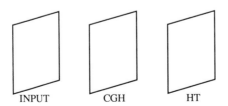

Figure 1.3 Optical Hough transform (HT) system using a CGH to implement multiple lens functions in a simple and cost-effective architecture.

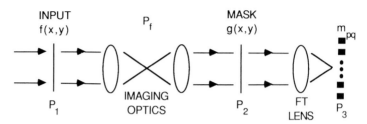

Figure 1.4 Optical system using frequency-multiplexed multiple filters or transfer functions achieving parallel multiplication and addition with multiple filters (Casasent, 1985).

calculate moments, but it is useful to demonstrate other unique functions and operations possible on optical systems. The input P_1 is imaged onto a mask at P_2 that contains N different 2-D functions $g_n(x, y)$ each on a different spatial frequency carrier. The light leaving P_2 is the point-by-point product of f and g_n (optical systems can perform parallel multiplications). The output FT lens sums each of these N 2-D products onto a different detector at P_3. Detector n has an output $\iint f(x, y) g_n(x, y) \, dx \, dy$. With the g_n being different monomial functions, the P_3 outputs are the moments m_{pq} of $f(x, y)$. Our present concern is the fact that an optical system (Figure 1.4) can use frequency-multiplexing (of N patterns at P_2). Thus, different paths through the system see different transfer functions and the processor can perform N different 2-D functions on the 2-D input data.

Well-engineered optical hardware versions of the WRD FT and HT systems have been fabricated and their use has been demonstrated. Thus, they are viable simple optical systems. The fact that no one currently widely markets them is the major reason why they have not seen wide use. The HT is particularly attractive, since from it one can produce the WRD FT, the 2-D FT, moments, etc. (Gindi & Gmitro, 1984).

1.4 Optical correlators for pattern recognition

The optical correlator architecture (Vander Lugt, 1964) is a very important one. A simplified version is shown in Figure 1.5. The 2-D input scene $f(x, y)$ is placed at P_1 on an SLM, its 2-D FT $F(u, v)$ is incident on P_2 where a matched spatial filter (MSF) $G^*(u, v)$ is recorded. The light leaving P_2 is the product FG^* and its FT formed at P_3 is the 2-D correlation $f \circledast g$. This operation is equivalent to shifting

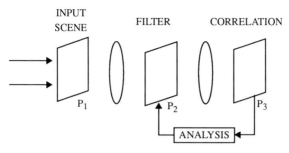

Figure 1.5 2-D optical correlator architecture (Casasent, 1992b).

the 2-D reference function *g* over all pixels in *f*. For each shift, the product of *g* and the corresponding region of *f* is formed and summed. The P_3 output contains a peak at locations in *f* where *g* is present. Thus, the architecture performs pattern recognition by template matching. The advantages of a correlator (optical or electronic) for locating objects are well known: it handles multiple objects in parallel and it provides the best detection in white Gaussian noise (due to the processing gain it provides). Figure 1.5 shows different filters fed to P_2 depending upon prior P_3 correlation results.

1.4.1 Fabrication

A major practical issue has always been whether one could fabricate a well-engineered and rugged optical correlator. Recent results have provided several such processors. Figure 1.6 shows a solid optics correlator that has been fabricated and tested (Lindberg & Hester,

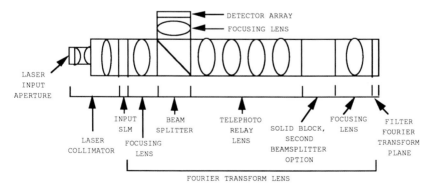

Figure 1.6 Solid optics correlator (Lindberg & Hester, 1990).

1990). It uses modular graded index solid optics to perform the FT and for magnification. The optics collimates a laser diode source at the left. This illuminates the input SLM, whose FT is incident on a reflective filter. The light then travels right to left through the same optics and from a beam-splitter onto an output detector array. Figure 1.7 shows another compact correlator presently being fabricated (Ross & Lambeth, 1991). It has all its components mounted around the perimeter of a circle. Light enters from the left, reflects off of the different elements (magneto-optic light-mod SLMs are used) with the correlation appearing on the charge-coupled device (CCD) detector. Chapter 5 discusses another compact correlator design in detail.

The present versions of the systems in Figures 1.5–1.7 are limited by the SLMs (at P_1 and P_2). Electrically-addressed SLMs in which the input image data are fed electrically from the sensor to P_1 (or data are fed electrically to P_2) are the most attractive approaches. Presently, the most available and reliable 2-D SLMs are binary devices. This significantly limits the use of these architectures. Promising gray-scale electrically-addressed devices may mature and make such architectures viable. As a result, much research attention has been given to techniques to design filters that can be recorded on SLMs with binary phase (± 1 transmittance values) and ternary ($0, \pm 1$ transmittance) values (Casasent & Chao, 1992). This work has yet to provide advanced distortion-invariant filters. However, considerable progress has been made. A binary device in the input plane is a more significant problem, since imagery is gray scale. Preprocessing methods to reduce

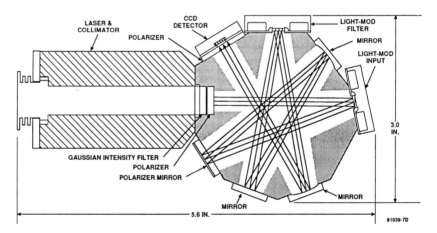

Figure 1.7 Hockey puck optical correlator (Ross & Lambeth, 1991).

a gray scale input to a binary pattern are conceptually possible, but are digital. It is unclear if one would analog-to-digital (A/D) convert the input, digitally preprocess it, reduce it to a binary pattern, and then perform optical correlations. By restricting the correlator to operate on binary inputs and binary filters, one would probably perform the correlation in digital hardware.

Section 1.5.1 details a viable preprocessing optical processor using binary devices. This algorithm and architecture are especially attractive, since the present optical SLMs can operate at 5000 frames/s data rates. In general, our remark (Section 1.2) that optical processors must employ analog data (to realize their full potential) is clearly true. We now address how a new algorithm and optical architecture emerged to address this issue (Psaltis, 1984). It avoids the need for 2-D SLMs by using other concepts and optical components, specifically acousto-optic (AO), that are analog and presently available. This is an excellent example of the need for and use of new algorithms and architectures with available operations and components to solve a major problem in optical processing.

1.4.2 2-D acousto-optic correlators

To describe the architecture, we first consider the 1-D optical processor in Figure 1.8. This is the time integrating (TI) AO correlator (Cohen, 1983). It consists of a light emitting diode (LED) or laser diode point source at P_1 fed with a 1-D signal $g(t)$. Lens L_0 collimates this source and uniformly illuminates P_2 with a signal $g(t)$ *with no spatial variation*. A 1-D AO device at P_2 is fed with a second signal $h(t)$, which travels down the AO cell producing a transmittance function $h(t - x/v_s)$ for P_2 that varies with time and space (distance x along P_2), where v_s is the velocity of sound in the AO cell. The light leaving P_2 is the product $g(t)h(t - x/v_s)$. This light distribution is

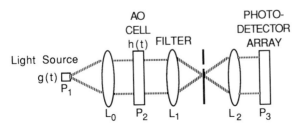

Figure 1.8 1-D TI AO correlator.

imaged (by L_1 and L_2) onto P_3. TI linear detector array at P_3 integrates
this light distribution in time yielding an output

$$\int g(t)h(t - x/v_s)\,dt = g \circledast t$$

that is the 1-D correlation of the 1-D signals g and t. We now consider
the final system in Figure 1.9. This is a multichannel version of Figure
1.8 with N input point modulators (LEDs or laser diodes) and N TI
linear 1-D detector arrays at P_3. The N 1-D outputs on N 1-D linear
detector arrays at P_3 are the correlations of $s(t)$ with the N reference
functions $r_n(t)$ fed to P_1. In the optical system, lenses L_1–L_3 image P_1
onto P_2 vertically and expand each P_1 output horizontally to illuminate
P_2 uniformly at a different angle of incidence. Plane P_2 is imaged
horizontally onto P_3 by lenses L_4 and L_6 and the N inputs at P_1 are
imaged vertically onto N rows at P_3.

To consider how Figure 1.9 implements a 2-D correlation, we feed
the 2-D reference function $r(x, y)$ to be recognized to the N inputs
$r_n(t)$ at P_1 (we assume an $N \times M$ reference function, with the M
samples for each row $r_n(t)$ of $r(x, y)$ fed time sequentially to the N
(number of image rows) inputs at P_1). This uses a 1-D input at P_1 (N
laser diodes or LEDs) and the space and time variables available to
input a 2-D function to the system. The first line of the input scene
$s(x, y)$ is fed to P_2 as $s(t)$. The 1-D correlations of it and the N 1-D
inputs $r_n(t)$ are formed on N rows on the 2-D TI detector at P_3. The
contents of P_3 are then shifted up by one row. The second line of the
input scene is fed to P_2 and the reference is repeated at P_1. The N 1-D
correlations of the N lines of the reference object and the second line
of the input scene are formed at P_3 and added to the prior (shifted) N
1-D correlations. This process is repeated for all rows of the input
scene. The data leaving the top of P_3 (one row at a time) is the 2-D
correlation of the 2-D reference function $r(x, y)$ (fed in time and space

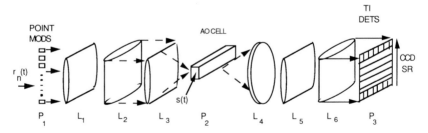

Figure 1.9 2-D TI AO correlator.

to P_1 as N 1-D functions $r_n(t)$) and the 2-D input scene $s(x, y)$ (fed one row at a time to P_2).

This optical architecture (Figure 1.9) is simple with no stringent alignment or positioning requirements. The optics all perform imaging (of P_1 onto P_2 and P_2 onto P_3) and no interferometric positioning tolerances are required. Thus, the optics are simple. The system uses available devices (1-D laser diode arrays, rugged 1-D AO devices and a 2-D CCD shift register (SR) detector) and can thus be fabricated with presently available components. All components are analog and thus a gray-scale correlation results. The reference function is easily adapted by feeding a new signal to the P_1 point modulators. The system can approach 1000 correlations per second (with appropriate multiplexing). The horizontal correlation is performed by time integration and the vertical correlation by shift and add time integration. This is an excellent example of our concept of the marriage of an analog processor and a new architecture with available components and operations for a specific problem.

Figure 1.10 shows the schematic and a photograph of a small, compact, rugged, fabricated and tested version of this new correlator architecture (Molley & Kast, 1992). Our purpose is not to detail the fabrication of the correlator in Figure 1.10 (or those in Figures 1.8 and 1.9), but rather to advance the key architectural and operational issues associated with each.

1.4.3 Distortion-invariant filters

Regardless of the analog optical correlator architecture used, the second major issue in such approaches is how to design a filter that will recognize all distorted versions of an object or a set of objects. An entire chapter or book could be devoted to this topic. Two conference volumes provide recent results (Casasent & Chao, 1992; Casasent & Chao, 1993). This section will only highlight these results, with emphasis on use of a hierarchical set of filters to achieve the desired results in realistic cases (considering full system details). These filters are of use in the coherent correlators (Figures 1.5–1.7) with gray-scale SLMs and in the non-coherent correlator of Figures 1.9 and 1.10. Our emphasis is again on how to use new distortion-invariant filters properly in a full scene analysis pattern recognition application when multiple objects of different classes are present in clutter. *These algorithmic efforts are*

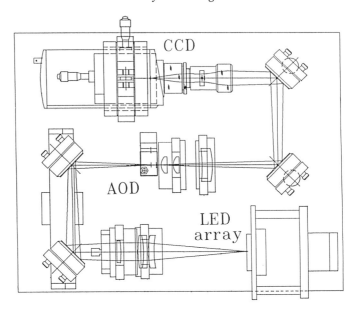

Figure 1.10 1-D AO pattern recognition correlator: (a) schematic; (b) photograph (Molley & Kast, 1992).

thus vital to make any optical correlator or pattern recognition system a useful element in a full system.

With the limitation of only a correlator architecture, one must thus devise intelligent filters that can extract the identity of each object in a scene. Reality dictates that one filter will not solve this problem. Rather, a sequence of filters is necessary. Each filter in this sequence has a different purpose and can thus be optimized for this goal. This is the algorithmic aspect of the approach emphasized in this section.

All approaches to 3-D distortion-invariant pattern recognition use a training set of views of each object in different distortions and form a filter that is a nonlinear combination of preprocessed versions of these training set images. The basic linear combination filter concepts (Hester & Casasent, 1980) have been significantly extended to include reduced sidelobes (Mahalanobis, Kumar, Vijaya & Casasent, 1987) and good performance in clutter (Ravichandran & Casasent, 1992). The newest (Casasent, 1992a) versions of these filters separate detection (recognizing all distorted versions of an object in the presence of clutter), false target rejection, and macro-class information (removing false alarms and providing target size and orientation macro-class information), followed by target identification. To achieve this, a sequence of three filters (for detection, recognition, and identification) is used. These are referred to as inference filters.

The set of training images in the FT domain is described by the columns of a data matrix $\mathbf{X}$. The filter H (in the frequency domain) is required to provide the same correlation peak value for all training images (these correlation peak values, typically '1', are the elements of a vector u). This first correlation peak constraint on H is described in linear algebra by

$$H^{\mathrm{T}}\mathbf{X} = u. \tag{1.1}$$

One can also minimize the signal energy E_s in the correlation plane,

$$E_s = H^+\mathbf{S}H, \tag{1.2}$$

where the diagonal matrix $\mathbf{S}$ has elements that are the envelope of the spectrum of the training set data. The solution to (1.1) and (1.2) produces a delta function correlation peak (the value at one point, the correlation peak, is constrained to unity, and all other correlation plane points are required to be a minimum). This localizes the correlation peak very precisely and hence provides an excellent estimate of the location of an object. It also reduces large sidelobes that can be present in the correlation plane. However, these filters have been shown to be poor in recognizing non-training set images. To address this, one defines a distortion spectrum **D** (a diagonal matrix whose elements are the maximum variation of the spectrum of the training set from the average). We also minimize its energy in the correlation plane,

$$E_d = H^+\mathbf{D}H. \qquad (1.3)$$

When clutter is present, we describe its spectrum by a diagonal matrix **N** and we minimize its energy in the correlation plane,

$$E_n = H^+\mathbf{N}H. \qquad (1.4)$$

The general form of the filter that satisfies these various constraints is

$$H = \mathbf{T}^{-1}\mathbf{X}(\mathbf{X}^+\mathbf{T}^{-1}\mathbf{X})^{-1}u, \qquad (1.5)$$

where the diagonal matrix **T** preprocesses the training set data. The elements of **T** are the maximum of several different energy terms that we wish to minimize, e.g. **T** = max {**S**, **N**} at each spatial frequency.

To provide reduced sidelobes and fewer false alarms due to clutter, we minimize both E_s and E_n. We minimize the energy $E = E_s + cE_n$, where the control parameter c is varied to emphasize which energy term it is of most concern to minimize. To provide large correlation peaks for all distorted versions of true-class objects to be detected, we use a large c value (this suppresses high spatial frequencies that differ for different distorted versions of an object). One could also minimize $E = E_s + cE_d$ to achieve this. To discriminate between several similar objects, we use a small c value (this emphasizes high spatial frequencies, which are those which distinguish similar objects).

Figure 1.11 shows an example of the use of a set of such filters with the control parameter c in $E = E_s + cE_n$ and the training set varied to achieve the different goals in the levels of computer vision. Figure 1.11(a) shows a scene with 13 different objects at different distortion angles in clutter (Figure 1.11(b) lists the objects, their orientations, and shows their distortions). The first filter is trained on only the small sized objects (ZSU-23) and a large $c = 0.1$ value is used. This filter recognizes different distorted versions of each object (due to the large c value) and it recognizes object regions of the scene equal to or larger than the smallest object. The correlation plane (Figure 1.11(c)) shows correlation peaks at the locations of all 13 true class objects and only one false alarm in the lower left. False alarms are allowed and expected in this first output, since they will be removed in subsequent levels. This output (Figure 1.11(c)) performs detection.

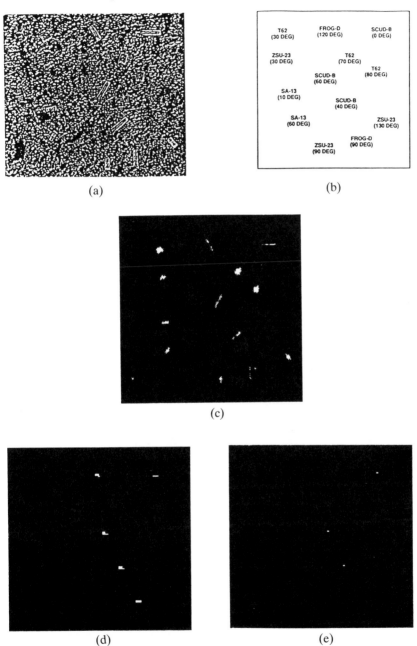

Figure 1.11 Example of the correlation results for a set of distortion-invariant inherence filters: (a) input; (b) object code; (c) detection; (d) macro-class recognition; (e) identification, (Casasent, 1992a).

With the locations of candidate objects known, we now apply a second filter intended to locate only the large objects in the scene (Frogs and Scuds). This filter uses a different training set (only the large objects) and a smaller $c = 0.005$ value. The output correlation plane has several false alarms; however, when we fuse its peaks with the locations of the detection peaks in Figure 1.11(c), only five correlation peaks occur (Figure 1.11(d)) at the locations of the five large objects. To perform the final identification of one object class (Scuds) and reject similar objects (Frogs), we formed a third filter trained only on Scuds with a still smaller $c = 0.001$ value (to provide discrimination). We fused its correlation plane with that in Figure 1.11(d) to obtain the result in Figure 1.11(e). This correlation plane contains three peaks at the locations of the three Scud objects and thus achieves object identification.

1.5 General purpose optical image processor

To be of full use in all aspects of one area such as image processing, an optical processor must perform more than simply distortion-invariant pattern recognition filters. Specifically, within image processing, there are a number of different general operations that are performed (e.g. morphological operations are performed in low-level vision, feature extraction is best for large-class problems and a classifier is needed). Optical systems (Section 1.3) can perform feature extraction. But all of these morphological feature extraction operations, plus new and attractive functions such as wavelet and Gabor transforms (Szu & Caulfield, 1992), should be implemented on the same optical architecture. This is necessary since a number of different optical architectures are not necessary, rather only one multifunctional general purpose architecture is needed (thus significantly reducing cost, size, etc.). This section addresses this issue.

As the single optical architecture we select a correlator, since it is shift-invariant and handles multiple objects in parallel (these properties are necessary in low-level vision), it has processing gain, and has a clear advantage (size, weight, power consumption, etc.) over electronic versions of the same architecture. To provide it with a more general capability, different filter functions are used at P_2 and different non-linear operations are employed on the P_3 correlation plane output in Figure 1.5. As the filter function at P_2 is varied, the general image processing operation performed can be selected and a general purpose multifunctional optical image processor results.

In general scene analysis and image processing, one can denote several different levels of machine vision with different functional purposes and using different operational techniques. One version of such a system (Figure 1.12) performs: clutter reduction followed by detection of candidate regions of interest (ROIs) in a scene where objects may exist; image enhancement of these ROIs; reduction of false alarms; extraction of macro-class information on the objects in each ROI; feature extraction; and finally classification of the identity of each object. The new algorithmic work required is to devise methods to implement these different operations with new filter functions so that all operations can be achieved on the same multifunctional general purpose optical correlator architecture of Figure 1.5.

1.5.1 Morphological optical processing

All major low-level computer vision operations can be described in image algebra as morphological operations (Haralick, Sternberg & Zhuang, 1987). The two basic operations in morphology are an erosion and a dilation. These involve, respectively, shrinking and growing *local image regions*. They are useful to remove noise and fill in holes in objects. They are typically used in pairs to perform an opening (erosion plus a dilation) or a closing (dilation plus an erosion). To implement these operations on a correlator, a structuring element (SE) filter is used at P_2 and the P_3 correlation plane is thresholded low (dilation) or high (erosion) (Casasent & Botha, 1988). The subtraction of an eroded and a dilated image produces an edge-enhanced image. Many other low-level image processing functions are also possible.

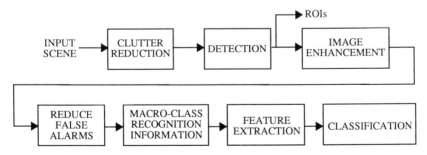

Figure 1.12 The levels of computer vision.

One algorithm (Casasent, 1992b) to use morphology for the clutter reduction function in Figure 1.12 is to open the input image, close the input image and form the difference of the closed and opened images. This removes large area background regions of different gray scale and makes all objects bright. It thus performs background normalization and object normalization. Opening and closing operations are also used for the large enhancement function in Figure 1.12 to improve each ROI before attempting identification (Casasent, 1992b). All of these clutter reduction and image enhancement operations are possible on the correlator of Figure 1.5 with different SE filters at P_2 and with low or high thresholds at P_3. The sizes of the SE filter determine the sizes of the background areas removed, the sizes of the holes filled in objects and the widths of the edges in the edge-enhanced images of each ROI. These are thus very basic low-level vision functions realizable with only several P_2 filter functions.

To implement these operations on gray-scale input imagery, one must separately non-linearly process binarized input images (thresholded at different gray levels) and then combine these separate binary morphological operations to produce the final gray-scale output morphologically processed scene. This is an ideal use for optical correlators with binary SLMs and high frame rates (Figure 1.7 and Section 1.4). The different binarized inputs are fed to P_1 in Figure 1.5, each binary (thresholded) morphological correlation is detected on an optically addressed SLM at P_3, and sequential P_3 outputs are read out and summed on a CCD detector array which contains the final desired result. The original version of this algorithm (Shih & Mitchell, 1989) intended for a pipelined systolic digital multiprocessor has been modified for an efficient optical realization (Schaefer & Casasent, 1992).

1.5.2 New undefined detection filters

To perform the object detection function in Figure 1.12, a suite of new P_2 filter functions was developed. These perform wavelet transforms (WTs) and Gabor transforms (GTs) on the input data to detect the presence of different-sized objects and different spatial frequencies in different orientations in local regions of different sizes in the input scene. These use new P_2 filter functions that are combinations of several WT or GT generating functions (Szu & Caulfield, 1992). They can provide image maps of clutter regions in the scene (Casasent, 1992b) and candidate object regions (Casasent, 1992b; Casasent,

Schaefer & Sturgill, 1992). The hit–miss transform (HMT) (Casasent, Schaefer & Sturgill, 1992) locates objects larger than some size S_1 and smaller than some size S_2. They achieve this by performing a foreground correlation on the object (with a hit filter) and a background correlation on the region around the object (with a miss filter). The final output is a non-linear combination of several non-linear operations and yields the locations of objects between size limits S_1 and S_2 set by the sizes of several SE filters at P_2. Space does not permit full details and examples of outputs from these different filter functions to be provided here. Such information is available in the references provided. Our present purpose is to note the growing number of new non-linear and general-purpose image processing operations on an optical correlator with different P_2 filter functions using new algorithms for the different purposes of the different levels of computer vision.

The presently preferable detection algorithm fuses the correlation plane outputs from HMT, WT and GT filters plus distortion-invariant filters to locate candidate object regions in the scene with a greatly reduced number of false alarms. Figure 1.13 shows one example of these results. The input (Figure 1.13(a)) contains six objects in severe clutter. We use the morphological clutter reduction algorithm and a WT filter to produce a clutter image of scene regions containing small clutter particles. We then fuse this output with one detection filter using GT functions and a second detection filter using the HMT that senses candidate object regions in the scene. The fusion algorithm uses

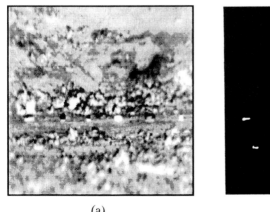
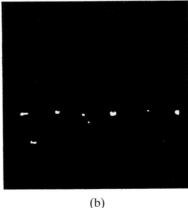

(a)   (b)

Figure 1.13 Clutter reduction and target detection by the fusion of several new morphological, WT/GT, and HMT operations on an optical correlator: (a) input; (b) output (Casasent, 1992b).

the fact that the clutter false alarms occur in different locations for the different algorithms. The final detection result (Figure 1.13(b)) is most impressive as all six objects are located with only two false alarms.

1.5.3 Reducing false alarms and macro-class information

The fusion operations noted in Section 1.5.2 and in Figure 1.11 reduce the number of false alarms and implement the operation noted in Figure 1.12. Figure 1.11(d) provided an example of extracting the macro-class information noted in Figure 1.12. This stage provides estimates of the size of the object in each ROI (other filters can provide estimates of the orientation of the object in each ROI). This macro-class information is of considerable use as it reduces the search space needed in the final classifier.

1.5.4 Feature extractor

With different filter functions (calculated using CGH techniques) input to P_2, the P_3 output will be different feature space values for the object region present in the P_1 input. All feature spaces noted in Section 1.3 can be optically produced on a correlator architecture with different P_2 filter functions. WT and FT features are also now possible. The locations of different peaks in the P_3 output now denote different features and the values of the different peaks denote the values for different features.

1.5.5 Fixed filter set architecture

The new algorithms advanced in Sections 1.4.3 and 1.5.1–1.5.4 allow many different general classes of operations to be performed with different filter functions: morphology, WT/GT filtering, distortion-invariance, etc. Thus, with such new algorithms, a correlator becomes a general-purpose image processor. We now note that there are realistically a modest number of different filter functions required for general image processing: several morphological SEs, several HMT filter sets, a few WT/GT filters for clutter and for different objects, several feature extraction filters, and a set of macro-class and distor-tion-invariant filters. We now advance an architecture that makes use of this reduced number of required image processing filters.

Figure 1.14 shows this architecture. It is a multichannel correlator

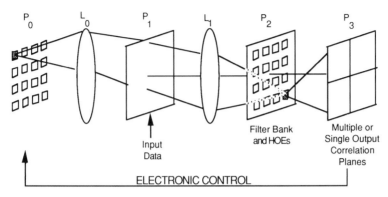

Figure 1.14 Programmable, general purpose, multifunctional optical image processor (Casasent, 1992b).

with the input scene at P_1, a set of multiple filters at P_2 and several (four) parallel correlation planes at P_3. The laser diode activated at P_0 selects one of the spatially-multiplexed filters at a different location in P_2. In each P_2 location, we can use the technique noted in Figure 1.4 to record several (four) frequency-multiplexed filters. Now, one laser diode at P_0 produces the correlation of the P_1 data with four different filter functions and four correlation patterns appear in P_3 in parallel. CGH techniques are used to record the filters in P_2 and each filter has an FT lens holographically recorded on it as a CGH to produce all correlations in the same spatial locations in P_3. We restrict the number of frequency-multiplexed filters to four since four correlation planes can be read out from the four different quadrants of a CCD detector array at P_3 and each correlation plane will have an adequate number of detection samples.

The use of space- and frequency-multiplexing with available laser diodes thus results in a new architecture. CGH techniques allow the filters and FT lenses to be recorded. The architecture is attractive since it avoids the need for an adaptive SLM at P_2, thus reducing system cost and complexity. This is thus another excellent example of how algorithms (for the different image processing functions), new correlation filters, available devices and components (CGHs and laser diode arrays), and a new architecture (with space- and frequency-multiplexed filters) are combined to provide a viable solution to a problem. The laser diode activation selects the filters used and thus the image processing function performed. Thus, a programmable, general-purpose, multifunctional optical image processor results.

1.6 Optical neural nets

Optical systems have unique properties that allow photons to cross with no interference or crosstalk occurring. This makes such systems very attractive for optical interconnections (Chapters 4 and 6). A neural net (NN) is one type of interconnection system that has potential use in our present optical image processing discussion. We now advance brief remarks on the use of NNs in image processing and the role for optics in these NNs.

Many NN architectures have been described for various image processing functions. For all low-level operations, correlation filters, and feature extraction operations, shift invariance is required; and, for the initial operations up to detection, multiple objects in the field-of-view must be handled and hence shift-invariance is necessary. All such NNs use iconic input neurons (one per image pixel) and hence require many $N_1 \approx 512^2$ input neurons. To provide shift-invariance, higher order NNs are possible, but they require $N_1^4 \approx 10^{23}$ interconnections, with more required for invariance to different distortions. Better techniques are available, specifically FT interconnections between the input neurons at P_1 and the weights at P_2 and between the weights and the output neurons at P_3 as shown in Figure 1.15. This shift-invariant NN is readily seen to be a correlator and is thus referred to as a correlation NN.

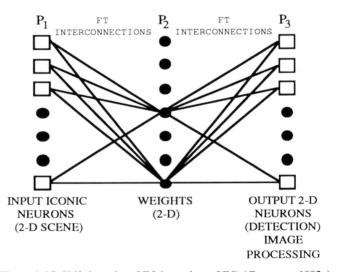

Figure 1.15 Shift-invariant NN (correlator NN) (Casasent, 1992c).

For all image processing functions up to the last two boxes in Figure 1.12, we recommend a correlator or a correlation NN (Casasent, 1992c) as in Figure 1.15. A correlation is, in fact, a powerful shift-invariant NN with up to 10^6 input neurons. All weights are applied in parallel to each local region of the input neuron array, where they multiply the input iconic neuron values and sum the resultant product into a correlation peak. For feature extraction, only one object is present and hence optical or electrical architectures can be used. However, the last box (the classifier) in Figure 1.12 is ideal for a classifier NN.

Classifier NNs are ideal for identification. These classifiers or optimization systems are unique, since they simultaneously calculate the discriminant functions used and the combinations of them used to produce decision surfaces to separate and classify input data into different classes. They are especially necessary when the number of object classes considered is large and when a wide range of distortions must be handled for each object class. In these cases, higher-order decision surfaces (not just piecewise linear ones) are necessary. For high storage capacity, analog input neurons (features) and analog weights are essential (Casasent, 1992c).

Both optical and electronic technology can be used to fabricate such neural nets. The on-line operation required in classifying input data is a set of two matrix–vector multiplications. In the training phase, a wider range of operation is required and thus optical NNs are significantly limited in the type of NN algorithms they can implement with existing components. When the input neurons to the NN classifier are features, the number of input neurons at P_1 is modest in size and digital techniques are presently sufficient to handle such cases. Thus, correlation NNs are clearly more practical at present, and there is more need for optical techniques in such NNs than in classifier NNs.

1.7 Summary and conclusion

These brief pages have provided many examples of the wide variety of operations possible on optical systems. These were shown to include a wide range of image processing functions: the calculation of many different feature spaces, all morphological low-level computer vision operations, new nonlinear WT/GT/HMT operations to reduce clutter and locate candidate object regions, and distortion-invariant filters to recognize objects. Clearly, the repertoire of operations (for image processing) possible on coherent optical systems is quite extensive. This was made possible by many new algorithm developments.

Many optical architectures exist. These can employ space- and frequency-multiplexing and allow very high density interconnections. In a correlator architecture, these interconnections become shift-invariant and different transfer functions exist for different filters or different paths through the system.

Optical hardware has matured in recent years. CGH techniques (Lee, 1992) allow many new operations to be implemented in very cost-effective systems. A number of compact and rugged optical correlator architectures have recently been fabricated. New algorithms have made use of available components such as laser diode arrays and AO cells to produce new architectures that are quite practical for 2-D correlations. Architectures employing laser diode arrays and integrated circuit CGH techniques with space- and frequency-multiplexing provide a 2-D correlator architecture that does not require an adaptive SLM filter. Recent advances in image processing algorithms have made a programmable general-purpose multifunctional optical image processor quite realistic.

This coherent optical image processing case study has shown how new algorithms can be used with available components to produce new optical architectures and an increasing suite of operations possible on viable optical processors.

References

Casasent, D. (1985). Computer generated holograms in pattern recognition: a review. *Optical Engineering*, **24**, 724–30.

Casasent, D. (1991). Optical processing and hybrid neural nets. In *Applications of Artificial Neural Networks*, S. K. Rogers, ed., Vol. 1469, *Proceedings Society Photo Optical Instrumentation Engineers*, Society of Photo Optical Instrumentation Engineers, Bellingham, WA.

Casasent, D. (1992a). An optical correlator, feature extractor, neural net system, *Optical Engineering*, **31**(5), 971–8.

Casasent, D. (1992b). Optical pattern recognition. In *Optical Technologies for Aerospace Sensing*, J. E. Pearson, ed., Vol. CR47, *Proceedings Society Photo Optical Instrumentation Engineeers*, Society of Photo Optical Instrumentation Engineers, Bellingham, WA.

Casasent, D. (1992c). Large capacity neural nets for scene analysis. In *Applications of Artificial Neural Networks III*, Vol. 1709, *Proceedings Society Photo Optical Instrumentation Engineers*, Society of Photo Optical Instrumentation Engineers, Bellingham, WA.

Casasent, D. and Botha, E. (1988). Optical symbolic substitution for morphological transformations, *Applied Optics*, **27**, 3806–10.

Casasent, D. and Chao, T. H., eds. (1992). *Optical Pattern Recognition IV*, Vol. 1701, *Proceedings Society Photo Optical Instrumentation Engineers*, Society of Photo Optical Instrumentation Engineers, Bellingham, WA.

Casasent, D. and Chao, T. H., eds. (1993). *Optical Pattern Recognition V*, Vol. 1959, *Proceedings Society Photo Optical Instrumentation Engineers*, Society of Photo Optical Instrumentation Engineers, Bellingham, WA.

Casasent, D., Schaefer, R. and Sturgill, R. (1992). Optical hit-miss morphological transform, *Applied Optics*, **31**, 6255–63.

Clark, D. and Casasent, D. (1988). Practical optical Fourier analysis for high-speed inspection, *Optical Engineering*, **27**(5), 365–71.

Cohen, J. D. (1983). Incoherent-light time-integrating processors. Chapter 9 in Berg, N. J. and Lee, J. N., *Acousto-Optic Signal Processing: Theory and Implementation*, M. Dekker, New York.

Fore, C. (1991). Radiation effects in optical processing systems, Defense Nuclear Agency Report DASIAC-91-1428, Defense Nuclear Agency, Alexandria, VA.

Gindi, G. R. and Gmitro, A. F. (1984). Optical feature extraction via the radon transform, *Optical Engineering*, **23**(5), 499–506.

Haralick, R., Sternberg, S. and Zhuang, X. (1987). Image analysis using mathematical morphology, *IEEE Patt. Anal. Mach. Intell.*, **9**, 532–50.

Hester, C. and Casasent, D. (1980). Multivariant technique for multi-class pattern recognition, *Applied Optics*, **19**, 1758–61.

Lee, S. H., ed. (1992). *Computer-Generated Holograms and Diffractive Optics*, Vol. MS33, *Proceedings Society Photo Optical Instrumentation Engineers*, Society of Photo Optical Instrumentation Engineers, Bellingham, WA.

Lendaris, G. G. and Stanley, G. L. (1979). Diffraction-pattern sampling for automatic pattern recognition. *Proc. IEEE*, **58**, 198–216.

Lindberg, P. C. and Hester, C. (1990). The challenge to demonstrate an optical pattern recognition system. In *Hybrid Image and Signal Processing II*, Vol. 1297, *Proceedings Society Photo Optical Instrumentation Engineers*, Society of Photo Optical Instrumentation Engineers, Bellingham, WA.

Liu, H., Davis, J. and Lilly, R. (1985). Optical data processing properties of a LCTV-SLM, *Optics Letters*, **10**, 635–7.

Mahalanobis, A., Kumar, B. V. K. Vijaya and Casasent, D. (1987). Minimum average correlation energy (MACE) filters, *Applied Optics*, **26**, 3633–40.

Molley, P. and Kast, B. (1992). Automatic target recognition and tracking using an acousto-optic image correlator, *Optical Engineering*, **31**(5), 956–62.

Psaltis, D. (1984). Two dimensional optical processing using one-dimensional input devices, *Proc. IEEE*, **72**, 962–74.

Ravichandran, G. and Casasent, D. (1992). Minimum noise and correlation energy (MINACE) optical correlation filter, *Applied Optics*, **31**, 1823–33.

Richards, J. and Casasent, D. (1989). Real-time optical Hough transform for industrial inspection. In *Intelligent Robots and Computer Vision VIII: Algorithms and Techniques*, D. Casasent, ed., Vol. 1192, *Proceedings Society Photo Optical Instrumentation Engineers*, Society of Photo Optical Instrumentation Engineers, Bellingham, WA.

Richards, J., Vermeulen, P., Barnard, E. and Casasent, D. (1988). Parallel holographic generation of multiple Hough transform slices, *Applied Optics*, **27**, 4540–5.

Ross, W. E. and Lambeth, D. N. (1991). Advanced magneto-optic spatial light modulator device development. In *Devices for Optical Processing*, D. M. Gookin, ed., Vol. 1562, *Proceedings Society Photo Optical Instrumentation Engineers*, Society of Photo Optical Instrumentation Engineers, Bellingham, WA.

Schaefer, R. and Casasent, D. (1992). Optical implementation of gray scale morphology. In *Nonlinear Image Processing III*, J. Astola, C. G. Boncelet and E. R. Dougherty, eds., Vol. 1658, *Proceedings Society Photo Optical Instrumentation Engineers*, Society of Photo Optical Instrumentation Engineers, Bellingham, WA.

Shih, F. Y. and Mitchell, O. R. (1989). Threshold decomposition of gray-scale morphology into binary morphology, *IEEE Patt. Anal. Mach. Intell.*, **11**(1), 31–42.

Szu, H. H. and Caulfield, H. J. (1992). Special section on wavelet transforms, *Optical Engineering*, **31**(9), 1823–16.

Vander Lugt, A. B. (1964). Signal detection by complex spatial filtering, *IEEE Trans. Inform. Theory*, **IT-10**(2), 139–45.

2

Noise issues in optical linear algebra processor design

STEPHEN G. BATSELL
Naval Research Laboratory

JOHN F. WALKUP and THOMAS F. KRILE
Department of Electrical Engineering, Texas Tech University

2.1 Introduction

Optical linear algebra processors (OLAPs) perform numerical calculations such as matrix–vector multiplication by equating a particular property of light, normally its intensity level, to a number. Through various effects to be discussed in this section, uncertainty is introduced either through random fluctuations or the addition of a bias. The fluctuating intensity can now be considered a random variable for which a mean and standard deviation can be determined. After corrections for bias, the mean value can be taken as the correct level that relates to the expected numerical value while the nonzero standard deviation represents processor noise.

In the design of analog OLAPs such as matrix–vector and matrix–matrix processors, the effect of noise from both device and system sources has a major impact on the choice of system architecture and components. The ability of a particular system design to meet system performance specifications such as signal-to-noise ratio (SNR), dynamic range, and accuracy is directly connected to the noise properties of the system and its components. As an example, the choice of the spatial light modulator (SLM) has a significant impact (Casasent & Ghosh, 1985; Perle & Casasent, 1990; Taylor & Casasent, 1986; Batsell, Jong, Walkup & Krile, 1989a, 1990; Jong, Walkup & Krile, 1986). This chapter examines the effects of noise on these system specifications and their implications for OLAP design.

We begin with a theoretical analysis of a simple matrix–vector multiplier and then generalize the results for N cascaded matrix–vector multipliers. From these analyses, the major noise sources will be identified and discussed. Their effects on the principal design criteria are then assessed. Finally, experimental results for systems employing several current SLMs are presented and discussed.

2.2 System considerations

Of the many linear algebra operations, this chapter focuses on optical processors that perform matrix–vector and matrix–matrix multiplication, as these are the most basic linear algebra operations (Caulfield & Rhodes, 1984). Researchers have investigated many architectures for optically performing these operations (McAulay, 1987; Goodman, Dias, Woody & Erickson, 1978; Goodman & Woody, 1977; Goodman, Dias & Woody, 1978; Dias, 1980; Athale, 1986; Athale & Collins, 1982). To simplify the analysis a generic matrix–vector multiplier which breaks the processor into five fundamental parts is used. These five parts are the input light source, the input propagation structure, the spatial light modulator, the output propagation structure, and the detector. Figure 2.1 shows this generic matrix–vector multiplier and indicates its operation mathematically. The primed variables represent electrical signal domain and the unprimed variables represent optical signal domain.

2.2.1 System analysis of an OLAP

There are two principal mathematical operations performed in an OLAP; addition and multiplication. Consider the addition of two random variables, x_j and c_j, where x_j is the light intensity of the jth channel and c_j is the light intensity due to background light or crosstalk from an adjacent channel. The mean value of the result, y_j, is given by

$$E(y_j) = E(x_j) + E(c_j). \tag{2.1}$$

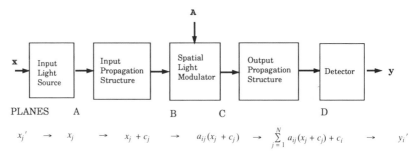

Figure 2.1 A generic matrix–vector multiplier.

The variance is given by

$$\sigma_{y_j}^2 = \sigma_{x_j}^2 + 2r_j\sigma_{x_j}\sigma_{c_j} + \sigma_{c_j}^2, \tag{2.2}$$

where r is the corresponding correlation coefficient. When the two random variables being added are uncorrelated, r_j is equal to zero and Eq. (2.2) becomes

$$\sigma_{y_j}^2 = \sigma_{x_j}^2 + \sigma_{c_j}^2. \tag{2.3}$$

While addition produces fairly straightforward results, the other operation, multiplication, has far more interesting results.

The underlining operation of interest in an optical matrix–vector multiplier is the multiplication of a vector $\mathbf{x}$ (input sources) by a matrix, $\mathbf{A}$ (SLM) transmittance or reflectance. The output of the operation, $\mathbf{y}$ (detected output) is given by

$$\mathbf{y} = \mathbf{Ax}. \tag{2.4}$$

Each element of the vector $\mathbf{x}$ is implemented by a light beam, while each element of the matrix $\mathbf{A}$ is represented by one or more cells of the SLM. In the presence of noise sources the elements of $\mathbf{x}$ and $\mathbf{A}$ are modeled as random variables. The fundamental principle in this operation is that the mean value of the light exiting the SLM is equal to the mean value of the incident light times the mean value of the reflection or transmission coefficient. This can be written as

$$E(\mathbf{y}) = E(\mathbf{A})E(\mathbf{x}). \tag{2.5}$$

The means and variances of the resulting matrix–vector product can now be determined. In this analysis the random variable x_j represents the jth element of the vector $\mathbf{x}$. Likewise, a_{ij} represents the ijth element of the matrix $\mathbf{A}$. As mentioned above the x_js will be incident light beams, while the a_{ij}s will be areas of the SLM. If the incident light, x_j, and the reflection/transmission coefficient, a_{ij}, are now considered random variables, it can be seen that for the above result to be true, the individual random variables must be uncorrelated. The assumption of the elements being uncorrelated is justified as long as the underlying noise process in the SLM is independent of the noise associated with the fluctuations in the incident light intensity. The random variables are further restricted to be widesense stationary or in the case of a periodic waveform, cyclostationary.

In Figure 2.1, the effects of noise sources on the random variables are shown. There are four principal planes of interest in the optical portion of the processor. Plane A is at the output of the light source where the input variable x_j' has been converted into a random light intensity, x_j. Plane B is at the input to the SLM. A random variable, c_j, represents all additional light incident on the ijth element of the SLM. This additional light is a combination of crosstalk and background light terms. Plane C is the output plane of the SLM. Finally, plane D is the input plane of the detector and c_i is a random variable representing the additional light incident on the ith detector. To simplify the following analysis, all crosstalk and background will be neglected and thus the c terms will be dropped. Their effects will be discussed in a later section. This simplified model is shown in Figure 2.2.

The mean and variance of the multiplication product, $a_{ij}x_j$, shown in Figure 2.2 will be determined. The mean light intensity is given by

$$E(a_{ij}x_j) = E(a_{ij})E(x_j). \qquad (2.6)$$

Similarly, the variance is given by

$$\sigma^2_{a_{ij}x_j} = E(a_{ij}^2)E(x_j^2) - [E(a_{ij})]^2[E(x_j)]^2, \qquad (2.7)$$

which can be rewritten as

$$\sigma^2_{a_{ij}x_j} = \{\sigma^2_{a_{ij}} + [E(a_{ij})]^2\}\{\sigma^2_{x_j} + [E(x_j)]^2\} - [E(a_{ij})]^2[E(x_j)]^2. \qquad (2.8)$$

This reduces to

$$\sigma^2_{a_{ij}x_j} = [E(a_{ij})]^2\sigma^2_{x_j} + [E(x_j)]^2\sigma^2_{a_{ij}} + \sigma^2_{a_{ij}}\sigma^2_{x_j}. \qquad (2.9)$$

From the above analysis, it can be seen that the same assumptions used in Eq. (2.2) to obtain the required multiplication relationship for the means have led to the variance relationship in Eq. (2.9). This relationship states that the output variance of the SLM will be composed of three terms. The first two terms state that the output variance will be signal-dependent, that is the mean values of x_j and a_{ij} act as multipliers which amplify or attenuate the corresponding variances depending on whether or not the mean values are greater or less than one. Furthermore, if the mean values fluctuate in time such as for temporally

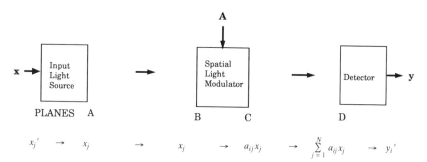

Figure 2.2 Simplified multiplication architecture.

modulated signals, the variance will likewise be temporally modulated. In the second term, if the mean value of the incident light is much greater than one, that has the effect of amplifying the SLM variance. Hence, small variance values of the SLM coefficients can have a major effect on the product variance of the product. Since the mean of the reflection or transmission coefficients is less than one for most current SLMs, the principal effect of signal-dependent noise will be temporal modulation. Finally, the third term is the product of the two variance terms as expected in a multiplication operation.

2.2.2 Cascaded multiplication operations

In the above analysis a single matrix–vector multiplication operation was considered. This result is now extended to the case of sequential multiplication operations, where they occur in cascade as shown in Figure 2.3. Again it is assumed that the incident light and the noise process of the kth SLM are uncorrelated for all k. Furthermore, it is assumed that each SLM has a statistically independent noise process with respect to any other SLM. The mean of a resulting product is given by

$$E(a_{ijk}x_{jk}) = E(a_{ijk})E(x_{jk}) \tag{2.10}$$

and the variance is given by

$$\sigma^2_{a_{ijk}x_{jk}} = E(a^2_{ijk})E(x^2_{jk}) - [E(a_{ijk})]^2[E(x_{jk})]^2, \tag{2.11}$$

which reduces to

$$\sigma^2_{a_{ijk}x_{jk}} = [E(a_{ijk})]^2\sigma^2_{x_{jk}} + \sigma^2_{a_{ijk}}\{[E(x_{jk})]^2 + \sigma^2_{x_{jk}}\}, \tag{2.12}$$

where k represents the kth SLM in the multiplication chain. When the subscript k is absent it is understood that k equals one as in the single multiplication case above. The term x_{jk} is a random variable that

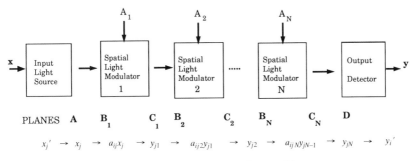

Figure 2.3 Cascaded multiplications architecture.

represents the incident light intensity on the kth SLM. For the case $k = 2$, the mean at plane B_2 is

$$E(y_{j1}) = E(a_{ij}x_j) = E(a_{ij})E(x_j). \qquad (2.13)$$

The mean at plane C_2 becomes

$$E(a_{ij2}y_{j1}) = E(a_{ij2})E(a_{ij1}x_j) = E(a_{ij2})E(a_{ij1})E(x_j), \qquad (2.14)$$

and the variance at plane C_k becomes

$$\sigma^2_{a_{ij2}y_{j1}} = \{\sigma^2_{a_{ij2}} + [E(a_{ij2})]^2\}\{\sigma^2_{a_{ij1}} + [E(a_{ij1})]^2\}\{\sigma^2_{x_j} + [E(x_j)]^2\}$$
$$- [E(a_{ij2})]^2[E(a_{ij1})]^2[E(x_j)]^2. \qquad (2.15)$$

For the case where there are N SLMs in the multiplication chain, the mean at plane C_N is given by

$$E(a_{ijN}y_{jN-1}) = \left[\prod_{k=1}^{N} E(a_{ijk})\right]E(x_j), \qquad (2.16)$$

and the variance at plane C_N by

$$\sigma^2_{a_{ijN}y_{jN-1}} = \left(\prod_{k=1}^{N} \{[E(a_{ijk})]^2 + \sigma^2_{a_{ijk}}\}\right)\{\sigma^2_{x_j} + [E(x_j)]^2\}$$
$$- \left\{\prod_{k=1}^{N} [E(a_{ijk})]^2\right\}[E(x_j)]^2. \qquad (2.17)$$

Hence, from Eq. (2.16) it can be seen that the effect of cascaded multiplication operations for the mean is that the resultant mean at plane C_N is simply the product of all the means. For the variance, on the other hand, each additional multiplication creates new cross product terms, thus increasing the noise in a nonlinear way.

2.3 Additional noise sources

In addition to the multiplication process discussed above, noise is generated in a number of processes that occur in the operation of an OLAP such as light generation, background light, crosstalk, and detection. These processes are reviewed using the generalized matrix–vector model (Figure 2.1) discussed previously.

2.3.1 Light source noise

Plane A in Figure 2.1 denotes the output of the light source. The light intensity at this point represents the vector, $\mathbf{x}$. The first order statistics of the light intensity at plane $\mathbf{A}$ are reviewed in this section.

Light can be categorized into two principal types, thermal and laser light, based on the means of generation. Thermal light is generated by spontaneous emission. This occurs when a group of excited atoms or molecules drops to a lower energy state in a random and independent manner emitting photons in this process. Examples of thermal light sources include the sun, light emitting diodes (LEDs), gas discharge tubes and incandescent bulbs. Thermal light can be further categorized by its polarization. For polarized thermal light, the instantaneous intensity obeys a negative exponential probability density function (Goodman, 1985, p. 124), that is,

$$p_I(I) = \begin{cases} \dfrac{1}{\sigma_I} \exp\left(-\dfrac{I}{\sigma_I}\right) & I \geqslant 0 \\ 0 & \text{otherwise.} \end{cases} \tag{2.18}$$

For polarized thermal light, the standard deviation is equal to the mean,

$$\sigma_I = E(I). \tag{2.19}$$

For unpolarized thermal light, the instantaneous intensity obeys the Rayleigh probability density function (Goodman, 1985, pp. 125–6)

$$p_I(I) = \begin{cases} \dfrac{2I}{\sigma_I^2} \exp\left(-\dfrac{\sqrt{2}I}{\sigma_I}\right) & I \geqslant 0 \\ 0 & \text{otherwise.} \end{cases} \tag{2.20}$$

For unpolarized thermal light, the mean intensity is related to its standard deviation by

$$E(I) = \sqrt{2}\sigma_I. \tag{2.21}$$

Laser light is generated by stimulated emission (Goodman, 1985, pp. 138–50). In this process again atoms or molecules are excited to a higher energy state except the process is controlled to allow only a discrete number of higher states. Likewise, the excited atoms or molecules are allowed to decay to only a discrete number of lower states. For a single mode laser, there is only one higher and one lower state, resulting ideally in a single amplitude and phase of the resulting light. When more than one mode of oscillation occurs, the result is called multimode laser light. While the principal emission process is stimulated emission, the spontaneous emission process is also occurring, since it is this process that is frequently used to trigger stimulated emission. Therefore, instead of the ideal result of a constant amplitude and phase light wave, the outgoing intensity is composed of a large stimulated emission, with constant amplitude and single phase and a smaller spontaneous emission with a randomly varying component. The resulting intensity approximately follows the Gaussian probability density function,

$$p_I(I) \cong \left\{ \frac{1}{\sqrt{(2\pi)}\sigma_I} \exp\left[-\frac{(I - I_S)^2}{2\sigma_I^2} \right] \right\}, \quad I_S \geqslant E(I)_N. \quad (2.22)$$

Here I_S represents the stimulated emission component and $E(I)$ represents the small spontaneous emission component. The mean and variance are related by

$$\sigma_I^2 = 2I_S E(I)_N. \quad (2.23)$$

When multiple modes exist, the resulting probability density function approaches that of polarized thermal light. For N multiple modes of equal strength, the ratio of the standard deviation to the mean intensity is given by

$$\frac{\sigma_I}{E(I)} = \sqrt{\left(1 - \frac{1}{N}\right)}. \quad (2.24)$$

Laser diodes often approximate the single mode quite closely, while a large gas laser approximates the multimode laser case. The careful choice of the light source to minimize noise is an important design consideration.

2.3.2 Background noise/crosstalk

In propagating between the elements of the OLAP, additional light can degrade the x_js. This degradation can result from two sources: background light and crosstalk. Background light can come from a variety of sources including inadequate blocking of light between the pixels of the SLM, the scattering of light from the face of the SLM, the walls or the cabinet housing the processor. Regardless of the original source of the background light, it can usually be considered thermal light due to the scattering process. Crosstalk is caused by light leaking from adjacent channels in the processor. It usually retains the properties of the light in the originating channel. Background light or crosstalk can be represented for the jth channel by a random variable c_j. At planes B and D, the variable c_j representing background light and/or crosstalk can be added to the x_js for the jth channel. The principal effect is to create a bias term. As the mean value of the light intensity at each plane represents a numerical value, a sufficient offset can increase the intensity such that it now is interpreted as another number leading to erroneous results. The increase in the variance can lead to a similar result.

2.3.3 Detector noise

The mean and variance of the detected signal for an incident intensity, I, when a p-i-n photodiode and transimpedance amplifier are used, was derived in (Batsell, 1990). Here the principal results are reviewed. The mean voltage is given by

$$E(V) = \left(\frac{2.0 \times 10^5 \eta e A}{h \bar{v}}\right) E(I) + E(V_B), \qquad (2.25)$$

where 2.0×10^5 is the transimpedance term relating voltage and current, η is the quantum efficiency of the detector, h is Planck's constant, v is the mean frequency of the incident light, e is the electron charge, A is the area of the detector, and $E(I)$ is the mean incident intensity. The mean bias, $E(V_B)$ is a combination of effects due to the dark current and the background light incident on the photodetector.

The variance of the photodiode output voltage can be shown to be given by

$$\sigma_v^2 = (2 \times 10^5)^2 \left[\frac{4 \eta e^2 A E(I)}{3 h E(\bar{v}) t_T} + \left(\frac{\eta e}{h E(\bar{v})}\right)^2 \sigma_I^2 + \sigma_{dk}^2\right] + \sigma_{th}^2, \qquad (2.26)$$

where t_T is the transit time for electrons in the p-i-n photodiode, $E(v)$ is the mean frequency of the incident light, σ_I^2 is the variance of the incident intensity, σ_{dk}^2 is the variance of the dark current, and σ_{th}^2 is the variance of the thermal noise process (voltage). The first term in the brackets to the right of the equal sign in Eq. (2.26) is a signal-dependent term caused by the discrete nature of the photodetection process (Goodman, 1985, pp. 466–70). The next term is the variance of the incident light times a constant and can be signal-dependent or signal-independent depending on the source used. Finally, the last two terms on the right side of the equation are signal-independent and are caused by thermal noise processes in the detector and amplifier. Hence, in general, the detected voltage noise can be divided into signal-dependent and signal-independent parts.

2.4 Design criteria

Given the noise sources described above, their impact on the design of OLAPs will now be assessed. The two questions to be answered in this section are (1) what are the effects of these noise sources on the design criteria of OLAPs and (2) how can careful selection of components minimize these effects?

2.4.1 SNR

The SNR is defined here as the ratio of the mean value of a random variable to its standard deviation. In an OLAP, a SNR can be determined for the light intensity at each of the four planes indicated in Figure 2.1 and for the output voltage. In the latter case, the SNR at the output of the OLAP is given by

$$\text{SNR}_{\text{OUT}} = \frac{E(V)}{\sigma_V} = \frac{CE(I) + E(V_B)}{\sqrt{[K_1 E(I) + K_2 \sigma_I^2 + K_3 \sigma_{\text{dk}}^2 + \sigma_{\text{th}}^2]}}, \quad (2.27)$$

where the constant terms in Eqs. (2.25) and (2.26) have been replaced by the constants C, K_1, K_2, and K_3. Different terms will dominate in the denominator as a function of the incident light intensity. The dark current and thermal noise variances are independent of the incident light except at very high intensities which effectively raise the temperature of the detector. By cooling the detector and amplifier these noise terms can be minimized and are only important at very low light levels. At high light levels, the variance of the incident light is the dominant noise term when signal-dependent noise is present (Taylor & Casasent, 1986). For this case minimizing the SNR at plane D will minimize the SNR at the processor output. It was earlier shown that the multiplication operation results in signal-dependent noise at the output of the SLM. It is therefore important to analyze the effects of the multiplication operation on the resulting SNR at plane C.

Noiseless SLM

First, consider the case of a noiseless SLM; that is when the variance of the SLM coefficient is equal to zero and only the first term to the right of the equal sign of Eq. (2.9) exists. The SNRs for plane B and plane C are then given by

$$\text{SNR}_B = \frac{E(x_j)}{\sigma_{x_j}} \quad (2.28)$$

which yields

$$\text{SNR}_C = \frac{E(a_{ij}x_j)}{\sigma_{a_{ij}x_j}} = \frac{E(a_{ij})E(x_j)}{E(a_{ij})\sigma_{x_j}} = \frac{E(x_j)}{\sigma_{x_j}} = \text{SNR}_B. \qquad (2.29)$$

Note that in Eq. (2.29), the first term in Eq. (2.9) has been substituted. For a noiseless SLM the multiplication process had no effect on the SNR, as expected. Hence a noiseless SLM only attenuates or amplifies the input intensity but adds no noise, as one would expect from a neutral density filter or noiseless optical amplifier.

Noisy SLM

Next, we consider the case of a very noisy SLM; that is where the second and third terms in Eq. (2.9) so dominate the first term that the first term can be ignored. These dominant terms are the excess noise introduced by randomness associated with the SLM. The second and third terms introduce a signal-dependent noise term regardless of the statistical nature of the incident illumination at plane B as $\sigma_{x_j}^2$ can also be signal-dependent. When this signal-dependent noise term dominates, the equation for the SNR at plane C becomes

$$\text{SNR}_C = \frac{E(a_{ij})E(x_j)}{\sigma_{a_{ij}}E(x_j)} = \frac{E(a_{ij})}{\sigma_{a_{ij}}} = \text{SNR}_{\text{SLM}}, \qquad (2.30)$$

which occurs when $\text{SNR}_B \gg 1$. When the signal-dependent noise term in Eq. (2.9) dominates, then SNR_C depends only on the SNR_{SLM} and is independent of the SNR_B. Next, consider the third term of Eq. (2.9) which is a product of the variances of a_{ij} and x_j. Note that the variance of the light at plane B may be signal-dependent or signal-independent in this equation. At plane C, the SNR becomes

$$\text{SNR}_C = \frac{E(a_{ij})E(x_j)}{\sigma_{a_{ij}}\sigma_{xj}} = \left[\frac{E(a_{ij})}{\sigma_{a_{ij}}}\right]\left[\frac{E(x_j)}{\sigma_{x_j}}\right] = (\text{SNR}_{\text{SLM}})(\text{SNR}_B). \qquad (2.31)$$

In this case, the resulting SNR_C is the product of SNR_B and SNR_{SLM}.

Up to this point, only the effect of the SLM on Eq. (2.9) has been considered. However, the type of incident light also plays a significant role. For this reason, three special cases for the system illumination will be analyzed to see how the variance of the multiplication product at plane C in Eq. (2.9) is affected. These cases are (1) laser light, (2) polarized thermal light, and (3) unpolarized thermal light.

Laser light

In the case of laser light, the incident intensity $p(x_j)$ at plane B can be modeled as a narrowband Gaussian process (Goodman, 1985, pp. 141–4). Therefore, the mean and variance of x_j are not explicitly related, and the variance of the light at plane C is as given in Eq. (2.9). As the variance of the light at plane B $(\sigma_{x_j}^2)$ decreases and an ideal laser model where the probability density function of the intensity is a Dirac delta function $(\sigma_{x_j}^2 = 0)$ is approached, the variance of the light at plane C given by Eq. (2.9) is dominated by the signal-dependent term $(E(x_j) \gg \sigma_{x_j})$. Conversely, as the variance of the light at plane B increases, exceeding the value of the square of the mean, the variance at plane C is dominated by the signal-independent term in Eq. (2.9) discussed previously. Hence, depending on which terms dominate, the variance at plane C can be signal-dependent or signal-independent.

Polarized thermal light

For the case of linearly polarized thermal light, the probability density distribution of the incident intensity at plane B can be modeled by a negative exponential distribution where $E(x_j) = \sigma_{x_j}$. In this case Eq. (2.9) becomes

$$\sigma_{a_{ij}x_j}^2 = [E(a_{ij})]^2[E(x_j)]^2 + 2\sigma_{a_{ij}}^2[E(x_j)]^2, \qquad (2.32)$$

where both terms in the variance are now signal-dependent. Using this result, SNR_C becomes

$$SNR_C = \frac{E(a_{ij})E(x_j)}{(\sqrt{\{[E(a_{ij})]^2 + 2\sigma_{a_{ij}}^2\}})E(x_j)} = \frac{SNR_{SLM}}{\sqrt{[(SNR_{SLM})^2 + 2]}}. \qquad (2.33)$$

There are two limiting cases to be considered. First, when $SNR_{SLM} \gg \sqrt{2}$, Eq. (2.33) approaches one as an upper limit. Substituting $E(x_j) = \sigma_{x_j}$ in Eq. (2.28), SNR_B is also one. Hence, this is just the ideal SLM case for linearly polarized thermal light. Next, when the variance of the SLM coefficient is greater than its mean value, SNR_C in Eq. (2.33) approaches $0.707\, SNR_{SLM}$.

Unpolarized thermal light

In the case of unpolarized thermal light, the probability density function of the incident intensity is given in Goodman (1985), where $\sigma_{x_j}^2 = E(x_j)^2/2$, and Eq. (2.9) becomes

$$\sigma_{a_{ij}x_j}^2 = E(a_{ij})^2 \frac{E(x_j)^2}{2} + 3\sigma_{a_{ij}}^2 \frac{E(x_j)^2}{2}. \qquad (2.34)$$

Again, both terms are signal-dependent. Using Eq. (2.34) in Eq. (2.33) SNR_C becomes

$$SNR_C = \frac{E(a_{ij})E(x_j)}{\{\sqrt{[E(a_{ij})^2 + 3\sigma_{a_{ij}}^2]}\}\dfrac{E(x_j)}{\sqrt{2}}} = \frac{\sqrt{2}E(a_{ij})}{\{\sqrt{[E(a_{ij})^2 + 3\sigma_{a_{ij}}^2]}\}}. \quad (2.35)$$

Here there are two limiting cases, the noiseless and noisy SLMs. For the noiseless SLM, SNR_C in Eq. (2.35) reduces to 1.414 which referring to Eq. (2.28) and the condition on the variance given above is identical to the SNR_B. Next, looking at the noisy SLM case, the SNR_C is 0.816 SNR_{SLM}. Again, this is not the result expected from Eq. (2.30), thereby showing the effect of the incident light at plane B on SNR_C.

It can be seen from the above analysis that for the noiseless SLM, regardless of the type of light incident, $SNR_C = SNR_B$. However, in the noisy SLM case, the limit approached by SNR_C is directly dependent on SNR_B. When $SNR_B \gg 1$, SNR_{SLM} is the limiting factor in the overall system performance as would be the case for the ideal laser. As the mean and standard deviation of the incident light approach each other the limit becomes something less than the SNR_{SLM} as can be seen in the unpolarized thermal light case. When the mean and standard deviation of the light at plane B are equal (as for polarized thermal light) the limit has reached 0.707 SNR_{SLM}. Finally, as the mean becomes much less than the standard deviation the limit becomes the product of the SNR_B and $SNR_{SLM} \div \sqrt{3}$.

We have shown that the multiplication process sets fundamental limits on SNR_C. Even when SNR_{SLM} is high, approaching the noiseless SLM case, the resulting SNR at plane C can never exceed SNR_B. On the other hand, when $SNR_{SLM} \ll 1$, the resulting $SNR_C \leqslant SNR_{SLM}$. This fundamental limit for the noisy SLM case also shows that regardless of the properties of the light incident at plane B, as SNR_B increases the resulting light at plane C becomes signal-dependent.

2.4.2 Dynamic range

The dynamic range (DR) of the system is defined as the ratio of the maximum signal level to the minimum signal level. The dynamic range can be measured at each of the four planes of Figure 2.2 and at the output of the OLAP. For example, at the system output the dynamic range is the ratio of the maximum to minimum mean voltage,

$$DR = \frac{E[(V_i)_{\text{Max}}]}{E[(V_i)_{\text{Min}}]}. \tag{2.36}$$

The dynamic range of the OLAP is determined by the smallest of these five dynamic ranges. Currently the limiting factor for the dynamic range is the SLM and the multiplication operation. The SLM limits the dynamic range by the contrast ratio available. In a typical two-dimensional SLM, individual pixels can be in one of two states, 'on' or 'off'. In the 'off' state the SLM ideally blocks all the light while in the 'on' state the SLM passes all the light. However, in practical systems some light passes through the SLM either through the pixel or between pixels yielding, in the 'off' state, a nonzero voltage output above the noise floor. In the 'on' state, the SLM attenuates the incident light somewhat reducing the recorded output voltage. The net effect is to reduce the contrast ratio and hence the dynamic range of the system. The multiplication operation further reduces the dynamic range because of the signal-dependent noise resulting from the operation. As the light intensity increases so does the standard deviation. At some point the standard deviation is too large to measure the signal level accurately thus reducing the usable maximum signal level of the system and the dynamic range.

2.4.3 Accuracy

As the goal of an OLAP is to represent a numerical value by some property of light, such as its intensity, and then to manipulate the light so as to perform a numerical operation, how accurately the light can be transformed back to a numerical value is of great importance. In order to compare the accuracy of OLAPs with competing digital electronic processors a meaningful definition of accuracy is required.

Definition of accuracy for a digital processor

Accuracy in a digital system is typically defined in terms of the achievable number of bits. In order to determine the accuracy of an analog signal in the same terms, an analog to digital (A/D) converter must be used as the interface between the analog and digital signal formats. The number of bits of accuracy of the output signal from an A/D converter is directly proportional to the number of distinguishable levels at the A/D converter input by the relationship that the number of levels equals 2^n. For example, a digital signal with eight bits of accuracy is obtainable from an analog signal divisible into 256 separable levels. These levels are taken to be uniformly spaced across the full scale range (FSR) of the system where the FSR is determined by the difference between the maximum signal level without saturation and the minimum detectable signal level in volts (Zuch, 1979; Gersho, 1978; Seifert & Nauda, 1989; Spenser, 1988). Furthermore, it is commonly assumed that the principal noise in the input signal at the A/D converter is signal-independent, additive, white, Gaussian noise.

With these assumptions in mind, the width of any quantization level can be defined as q, where

$$q = \text{FSR}/2^n, \tag{2.37}$$

and n is the number of bits of accuracy desired. Furthermore, the accuracy in bits, n, can be determined in terms of the resolution by

$$n = \log_2\left(\frac{\text{FSR}}{q}\right). \tag{2.38}$$

If q is set equal to $m\sigma_V$, where σ_V is the standard deviation of the voltage at the A/D converter input and m is a scale factor (an integer), the accuracy, n, can defined in terms of the noise standard deviation, σ_V, by

$$n = \log_2\left(\frac{\text{FSR}}{m\sigma_V}\right). \tag{2.39}$$

For the case of signal-independent noise, and hence uniform quantization levels, Bayesian detection theory can be used to set m such that the probability that a signal falls within a corresponding interval q is arbitrarily close to unity (Whalen, 1971; Van Trees, 1968). As an example, consider a Gaussian signal where m is set at 6 with $(m/2)\sigma$ on each side of the mean. In this case, there is over a 99% probability that the signal falls within the interval, q.

Accuracy defined for an optical processor

The above definition for accuracy in terms of the number of bits can be applied to an optical processor using the model previously derived (Batsell *et al.*, 1990). Since the FSR is given as the difference between the maximum usable signal and the minimum signal level, it is possible to define the FSR for the ith output channel as

$$\text{FSR}_i = E[(V_i)_{\text{Max}}] - E(V_b), \tag{2.40}$$

where $E[(V_i)_{\text{Max}}]$ is the mean maximum usable voltage at the output of the ith channel of the optical processor and $E(V_b)$, the bias, is the mean output noise voltage due to the combination of the dark current and any scattered light due to crosstalk or background light. FSR can then be rewritten as

$$\text{FSR}_i = C \sum_{j=1}^{N} E(a_{ij}) E[(x_j)_{\text{Max}}], \qquad (2.41)$$

where we have substituted for $E[(V_i)_{\text{max}}]$ the following equation,

$$E[(V_i)_{\text{Max}}] = C \sum_{j=1}^{N} E(a_{ij}) E[(x_j)_{\text{Max}}] + E(V_b). \qquad (2.42)$$

Here C represents constant terms described previously, $E[a_{ij}]$ is the mean of the SLM (transmittance or reflectance), and $E[(x_j)_{\text{Max}}]$ is the mean of the incident light at plane B for maximum usable intensity. This result can then be substituted into Eq. (2.39) to obtain the accuracy for the ith output channel as

$$n_i = \log_2 \left\{ \frac{C\sum_{j=1}^{N} E(a_{ij}) E[(x_j)_{\text{Max}}]}{m\sigma_{V_i}} \right\}, \qquad (2.43)$$

where σ_{V_i} is the standard deviation of the detected output voltage for the ith channel. Finally, Eq. (2.9) is substituted for the standard deviation, and

$$n_i = \log_2 \left\{ \frac{C\sum_{j=1}^{N} E(a_{ij}) E[(x_j)_{\text{Max}}]}{m\sqrt{\{K_1\sum_{j=1}^{N} E(a_{ij}) E[(x_j)_{\text{Max}}] + K_2\sigma_{I_i}^2 + K_3\sigma_{\text{dk}}^2 + \sigma_{\text{th}}^2\}}} \right\}, \qquad (2.44)$$

where the first term under the square root represents the photon shot noise, the second term represents the variance of the incident light at plane D in Figure 2.2, the third term represents the dark current in the detector, and the fourth term represents the thermal (Johnson) noise due to the detector/amplifier. The K_is represent constant terms shown in Eq. (2.27). The above equation relates the digital concept of accuracy in terms of the number of bits with the general statistical model for optical linear algebra processors derived earlier (Batsell *et al.*, 1990).

Upper and lower bounds on accuracy

Unlike an electronic system where signal-independent Gaussian noise is commonly a valid assumption, an optical processor's output often contains signal-dependent noise (Goodman, 1985). Therefore, uniformly spaced decision levels will not, in general, produce the maximum accuracy possible. Nonuniform spacing of the decision levels is necessary to produce optimum accuracy, i.e., the maximum number of analog levels at the quantizer. In practice, nonuniform threshold spacing based on the changing standard deviation of the signal is difficult to realize. Considerable insight can be gained by setting maximum and minimum bounds for the accuracy of an optical processor based on uniform quantization. Consider the case where the levels are spaced by $m\sigma_V$ and σ_V increases with signal voltage as would be the case in signal-dependent noise (see Figure 2.4). The levels will then be narrowest at the minimum signal voltage and widest at the maximum signal voltage. It will therefore be possible to set a minimum accuracy bound by considering the uniformly sampled case using the maximum voltage. Similarly, a maximum accuracy bound can be determined by using the uniformly sampled case based on the minimum voltage. In so doing, fundamental limits that bound the achievable accuracy of the system can be determined and these limits will remain true for a sampled analog system regardless of the type of quantization chosen.

A Accuracy upper bound The accuracy upper bound occurs when the maximum number of levels available are at the input to the quantizer. As the FSR of the system is constant, this occurs when the quantization bins with the minimum width are employed. The minimum width bin is determined by the noise floor at the detector output, as this

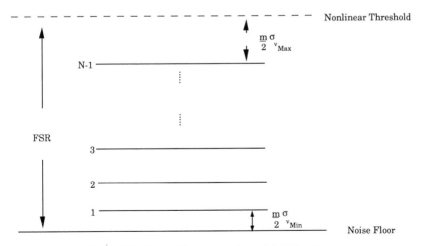

Figure 2.4 Nonuniform sampling with N levels.

corresponds to the minimum noise level for signal-dependent noise. In Eq. (2.44) and Figure 2.4, the noise floor is formed by the dark current and thermal noise. We thus obtain

$$n_{\text{Max}} = \log_2 \left\{ \frac{C\sum_{j=1}^{N} E(a_{ij}) E[(x_j)_{\text{Max}}]}{m\sqrt{(K_3\sigma_{\text{dk}}^2 + \sigma_{\text{th}}^2)}} \right\}. \tag{2.45}$$

Note that for this upper bound, the noise sources are signal-independent so that the maximum accuracy occurs in the signal-independent noise limit. Furthermore, this bound can only be improved by lowering the noise floor which is accomplished by cooling the detector, as both the dark current and thermal noise levels are related to temperature.

For those systems in which the detector has been sufficiently cooled, the photon shot noise process will determine the noise floor, and the bound will become

$$n_{\text{Max}} = \log_2 \left(\frac{C\sqrt{\{\sum_{j=1}^{N} E(a_{ij}) E[(x_j)]_{\text{Max}}\}}}{m\sqrt{K_1}} \right). \tag{2.46}$$

B Accuracy lower bound   The accuracy lower bound occurs at the maximum detected light intensity. As the maximum intensity level can vary between different processors depending on both the components and architecture used, the dominant noise source which determines the maximum level width can vary between processors. All noise sources can dominate, each for a different intensity range. Therefore, each case must be considered separately.

At the low light levels, the photon shot noise dominates and the minimum bound is equivalent to the upper bound. Increasing the intensity further, the variance of the incident light becomes the dominant noise process and the minimum accuracy bound is given by

$$n_{\text{min}} = \log_2 \left\{ \frac{C\sum_{j=1}^{N} E(a_{ij}) E[(x_j)_{\text{Max}}]}{m\sqrt{(K_2\sigma_{I_i}^2)}} \right\}. \tag{2.47}$$

As it has been shown that the variance of the light incident at the detector (plane D in Figure 2.2) is given by Eq. (2.9), this equation is substituted for $\sigma_{I_i}^2$ in Eq. (2.47), resulting in

$$n_{\text{Min}} = \log_2 \left[\frac{C\sum_{j=1}^{N} E(a_{ij}) E[(x_j)]_{\text{Max}}}{m\sqrt{(K_2\sum_{j=1}^{N}\{E(a_{ij})^2\sigma_{x_j}^2 + \sigma_{a_{ij}}^2(E[(x_j)]_{\text{Max}}^2 + \sigma_{x_j}^2)\})}} \right]. \tag{2.48}$$

Eq. (2.48) can be further simplified based on the noise properties of the SLM used and on the noise properties of the type of light generated at the light source. For this reason, two extreme cases will be considered; the noiseless SLM, where $\sigma_{a_{ij}}^2 = 1$, and the noisy SLM, where $\sigma_{a_{ij}}^2 > E(a_{ij})$.

1. Noiseless SLM For a noiseless SLM Eq. (2.48) reduces to

$$n_{\mathrm{Min}} = \log_2 \left(\frac{C \sum_{j=1}^N E(a_{ij}) E[(x_j)_{\mathrm{Max}}]}{m \sqrt{\{K_2 \sum_{j=1}^N [E[(a_{ij})]^2 \sigma_{x_j}^2]\}}} \right). \qquad (2.49)$$

For the case where the mean values of the SLM coefficients are constant over the array (or for a single element) this reduces to

$$n_{\mathrm{Min}} = \log_2 \left[\frac{CN(\mathrm{SNR}_{x\mathrm{Max}})}{m \sqrt{K_2}} \right] \qquad (2.50)$$

where $\mathrm{SNR}_{x\mathrm{Max}}$ is the SNR for a single input element for the maximum input intensity. Clearly, $\mathrm{SNR}_{x\mathrm{Max}}$ is the determining factor in Eq. (2.50) as all the other terms are constants.

Given this dependence on $\mathrm{SNR}_{x\mathrm{Max}}$, let us consider the following cases where the incident light is either a laser or thermal light.

(a) Ideal laser sources: For the ideal laser, where the mean is much greater than the standard deviation and hence there is a high $\mathrm{SNR}_{x\mathrm{Max}}$, we can achieve very good accuracy. Conversely, in the case of a noisy laser (one in which the SNR is low) the achievable accuracy is also very low.

(b) Thermal sources: For polarized and unpolarized thermal light, the standard deviation is equal to the mean ($\mathrm{SNR} = 1.0$) or equals the mean divided by the square root of two ($\mathrm{SNR}_x = 1.414$), respectively. Replacing the standard deviation, σ, by these values, the accuracy bound becomes, for polarized thermal light,

$$n_{\mathrm{Min}} = \log_2 \left[\frac{C \sum_{j=1}^N E(a_{ij}) E[(x_j)_{\mathrm{Max}}]}{m \sqrt{(K_2 \sum_{j=1}^N \{E[(a_{ij})]^2 E[(x_j)_{\mathrm{Max}}]^2\})}} \right] \qquad (2.51)$$

and for unpolarized light,

$$n_{\mathrm{Min}} = \log_2 \left[\frac{C \sum_{j=1}^N E(a_{ij}) E[(x_j)_{\mathrm{Max}}]}{m \sqrt{(2K_2 \sum_{j=1}^N \{E[(a_{ij})]^2 E[(x_j)_{\mathrm{Max}}]^2\})}} \right]. \qquad (2.52)$$

When the mean SLM coefficient is constant over the array (or for a single element) these equations reduce to

$$n_{\mathrm{min}} = \log_2 \left[\frac{C \sqrt{N}}{m \sqrt{K_2}} \right]^{(\mathrm{polarized})} \qquad (2.53)$$

and for unpolarized thermal light,

$$n_{\text{Min}} = \log_2 \left[\frac{C\sqrt{N}}{m\sqrt{(2K_2)}} \right]^{\text{(unpolarized)}}, \qquad (2.54)$$

where N is the total number of channels combined. For a single element case N equals one. From the last two equations it can be seen that for an ideal SLM (with constant mean value over N elements) to be combined with thermal light incident on the SLM, the minimum accuracy bound is a constant. This means that regardless of the size of the dynamic range used the lower limit on accuracy is the same as long as the noise in the incident light dominates the other source of noise.

2. Noisy SLM Next, we consider the case of a noisy SLM. Here the second term in the denominator of Eq. (2.48) is assumed to dominate and we can rewrite Eq. (2.48) as

$$n_{\text{Min}} = \log_2 \left[\frac{C\sum_{j=1}^{N} E(a_{ij})\,E[(x_j)]_{\text{Max}}}{m\sqrt{(K_2\sum_{j=1}^{N}\{\sigma_{a_{ij}}^2(E[(x_j)]_{\text{Max}}^2 + \sigma_{x_j}^2)\})}} \right]. \qquad (2.55)$$

Again for the case where all of the channels are identical this reduces to

$$n_{\text{Min}} = \log_2 \left(\left(\frac{C\sqrt{N}}{m\sqrt{K_2}} \right)(\text{SNR}_{\text{SLM}})\left\{ \frac{\text{SNR}_{x\text{Max}}}{\sqrt{[(\text{SNR}_{x\text{Max}})^2 + 1]}} \right\} \right) \qquad (2.56)$$

where SNR_{SLM} is the SNR of a single SLM channel. The accuracy equation is now subdivided into the product of three ratios indicated by the three sets of parentheses (i.e. a constant term, the SNR of the SLM, and a composite term). We will consider three possible cases for the composite term. First, when $\text{SNR}_{x\text{Max}} \gg 1$, the last term approaches one. This reduces Eq. (2.56) to

$$n_{\text{Min}} = \log_2 \left\{ \left[\frac{C}{m}\sqrt{\left(\frac{N}{K_2}\right)} \right]\text{SNR}_{\text{SLM}} \right\}. \qquad (2.57)$$

Here, we find the minimum accuracy limit is determined by SNR_{SLM}. Next, when $\text{SNR}_{x\text{Max}} \ll 1$, the accuracy limit becomes

$$n_{\text{Min}} = \log_2 \left[\left(\frac{C}{m} \right)\sqrt{\left(\frac{N}{K_2}\right)}(\text{SNR}_{\text{SLM}})(\text{SNR}_{x\text{Max}}) \right]. \qquad (2.58)$$

Finally the $\text{SNR}_{x\text{Max}}$ can be close to one. In particular, for thermal light, $\text{SNR}_{x\text{Max}}$ equals 1.0 or 0.707 depending on whether the light is polarized or unpolarized. The minimum accuracy limit then becomes

$$n_{\text{Min}} = \log_2 \left[\left(\frac{C}{m} \right) \sqrt{\left(\frac{N}{K_2} \right)} \left(\frac{1}{\sqrt{2}} \right) (\text{SNR}_{\text{SLM}}) \right]^{(\text{polarized})} \qquad (2.60)$$

or

$$n_{\text{Min}} = \log_2 \left[\left(\frac{C}{m} \right) \sqrt{\left(\frac{N}{K_2} \right)} \left(\frac{1}{\sqrt{3}} \right) (\text{SNR}_{\text{SLM}}) \right]^{(\text{unpolarized})}. \qquad (2.61)$$

Again, we find the accuracy is limited by the SNR_{SLM}. Clearly for a noisy SLM we find that the minimum accuracy limit is dominated by SNR_{SLM} and the type of light used in the OLAP is of secondary importance as long as its SNR_x isn't lower than that of the SLM. However, given that a particular SLM must be used in an architecture then the type of light chosen can greatly affect the accuracy.

2.4.4 Light efficiency

In designing an optical processor, the efficient use of light is important for minimizing power, size, and weight, as well as maintaining the system's dynamic range through the use of multiple SLMs cascaded together. The percentage of light attenuated in a SLM can be defined as

$$\% \text{ attenuation} = \frac{E[x] - E[ax]}{E[x]}, \qquad (2.62)$$

where $E[\ ]$ denotes an expectation value. In an ideal binary SLM, two states exist for a pixel: 100% transmission is desired in the 'on' state and 0% transmission is desired in the 'off' state corresponding to an a_{ij} equaling 1 or 0 respectively. In available commercial SLMs, less than 100% of the light is transmitted with the rest being either scattered or absorbed by the SLM. Likewise, in the 'off' state some light is transmitted through the pixel. For the case of $a > 1$, Eq. (2.62) would yield a negative percentage attenuation which would be defined as percentage gain.

A second measure of light efficiency is the contrast ratio (CR) for a given incident light intensity. The contrast ratio can be defined as

$$\text{CR} = \frac{E[a_{\text{on}}x]}{E[a_{\text{off}}x]}, \qquad (2.63)$$

where a_{on} is the pixel in the 'on' state and a_{off} is the pixel in the 'off' state. For area modulation, the contrast ratio determines the separation between adjacent mean value bins in the processor and the total number of bins available. As mentioned previously, the contrast ratio affects the dynamic range of the processor but differs from the dynamic range in that it is defined for a particular incident light intensity while the dynamic range is defined over all intensities.

2.5 Applications to some common SLMs

Up to now, analytical results have been presented based on the generalized OLAP model in Figure 2.1. In this section, the results of experimental measurements using several current SLMs and light sources are addressed. The section begins with an overview of the experimental procedure used followed by a description of the components used. Finally, each SLM used is discussed and experimental results are presented.

2.5.1 Experimental technique

As the generalized model is described here in terms of second order statistics, it is the mean and variance that need to be measured at each plane in the OLAP. However, it is not possible to measure the mean and variance of the light intensity directly. Rather, the light intensity must be detected and converted to an electrical voltage on which our equipment can operate.

Figure 2.5 is an overview of the experimental setup used for determining the mean and variance of the output voltage from the detector. A light source, SLM, and detector, along with the appropriate optics, made up the optical processor. Within the processor, numerical values were represented as intensity levels of the light. At the detector, the intensity level was transformed to an amplitude level of the output photocurrent of the detector. This was then transformed from the amplitude of a photocurrent to a voltage amplitude using a transimpedance amplifier. A multichannel analyzer then sampled this voltage and stored the number of times a particular amplitude occurred in the corresponding memory channel. The result of this process for a particular sampling time interval formed a histogram which could then be related to the probability density function of the signal. The sample mean and a sample variance were then determined from the appropriately normalized histograms. Each of the components used are discussed in more detail below.

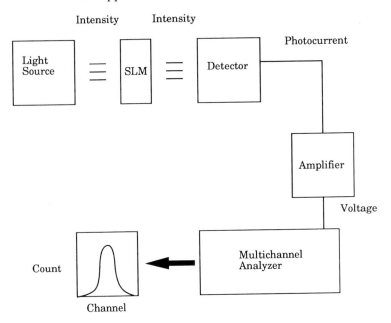

Figure 2.5 A conceptual block diagram of the experimental setup.

Light sources

Four light sources were used. Each had different noise properties. Two were thermal light sources (an LED and a tungsten–halogen lamp) and two were laser sources (a laser diode and a helium–neon gas laser). In addition, experimental plots of the standard deviation versus the mean and the SNR are included for each light source. For the tungsten–halogen lamp, the recorded histogram for three different source voltages is also included.

The LEDs used were Panasonic ultra-bright red LEDs which operate at 665 nm (Panasonic Data Sheet, 1988). The LEDs were driven by a DC power supply where the drive voltage at the supply could be varied between 0 and 20 V. However, the range was limited in the experiments to between 0 and 10 V to prevent the LED from burning out. A 330 Ω resistor was placed in series with the LED to regulate the current and protect the diode. Figure 2.6 is a plot of the standard deviation versus the mean voltage. Figure 2.7 is a plot of the SNR versus the power supply voltage, where the input voltage is proportional to the output light intensity of the LED. The SNR plot shows an

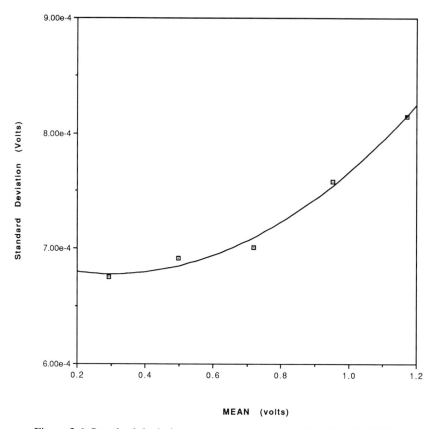

Figure 2.6 Standard deviation versus mean output voltage for the LED.

increasing SNR with intensity. Figure 2.6 shows the standard deviation also increases with increasing intensity. While the LED does generate light with a signal-dependent standard deviation as expected for a thermal source, it does not exhibit the constant SNR expected of a thermal source. This is due to the solid state cavity structure of an LED which restricts the light generated to a narrow frequency band. The resulting light behaves more like a multimode laser source with properties between those of a laser and a thermal source.

The white light source used was a tungsten–halogen lamp controlled by a stabilized power supply. The voltage to the lamp from the power supply could be set between 0 and 15 V by a control potentiometer, as well as be monitored by a meter. A polarizer was used to guarantee polarized thermal light. Figure 2.8 is a plot of the standard deviation

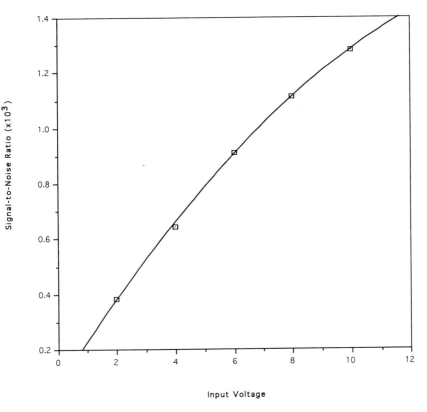

Figure 2.7 SNR for the LED.

versus the mean voltage. Figure 2.9 is a plot of the SNR versus the power supply voltage where the input voltage is proportional to the output light intensity of the tungsten–halogen lamp. Figure 2.8 shows the standard deviation to be signal-dependent and Figure 2.9 shows the SNR to be essentially a constant as expected for a thermal source. Figure 2.10 contains the histogram of the light source for three different input voltages. This is included as a reference for later histogram plots for each of the active SLMs showing temporal modulation effects. It should be clear for Figure 2.10 that for the tungsten–halogen lamp without an SLM present no temporal modulation effects are evident.

The helium–neon laser used was a Spectra-Physics Model 124B rated at 15 mW. The wavelength was at 632.8 nm and the laser operated in the TEM_{00} mode. The laser had a linearly polarized beam.

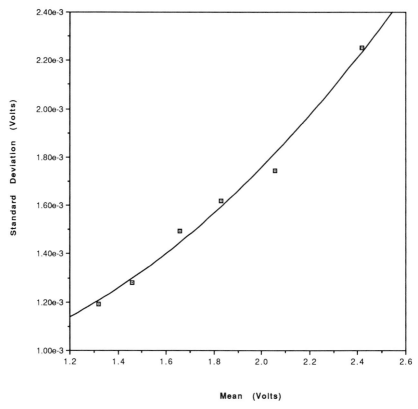

Figure 2.8 Standard deviation versus mean output voltage for the tungsten–halogen lamp.

The laser had a self-contained power supply and used a 115 V power source. A plot of the standard deviation versus mean output voltage is given in Figure 2.11 and a plot of the SNR versus light intensity level at the detector in Figure 2.12. As the helium–neon laser had no input voltage control to control the light intensity, the light intensity incident on the detector was changed by the use of neutral density filters. Figure 2.11 shows the standard deviation of the output voltage of the detector to be signal-dependent, while Figure 2.12 shows the SNR to be near constant.

The laser diode was a Sharp model LT020MC and it was controlled by an IR3C01 laser diode driver integrated circuit (IC). This laser diode is constructed using a GaAlAs double heterojunction in a V-channel substrate. The laser diode had built-in p-i-n photodiodes for

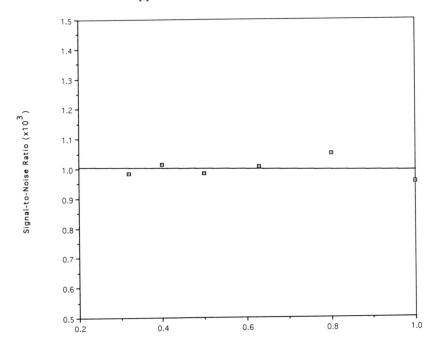

Figure 2.9 SNR for the tungsten–halogen lamp.

use in a feedback control circuit used to regulate its output power. It operated at 5 mW of power between 770 and 790 nm (Laser Diode User's Manual, 1986). Figure 2.13 is a plot of the standard deviation versus the mean output voltage. Figure 2.14 is a plot of the SNR versus the light intensity level at the detector. The laser diode's output light intensity was modified by the use of neutral density filters. Figure 2.13 shows that the standard deviation of the output voltage of the detector was signal-independent, while Figure 2.14 shows the SNR increased with increasing intensity.

SLMs

Two types of SLM exist; passive and active. In a passive SLM such as film or a neutral density filter, once the SLM is implemented in the system the amount of attenuation is set and cannot be changed except by removing the filter. In an active SLM the amount of attenuation can be changed either electronically or optically. Many active devices are

(a)

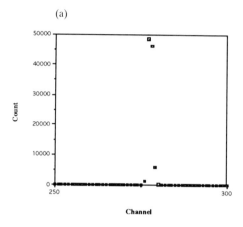

(b)

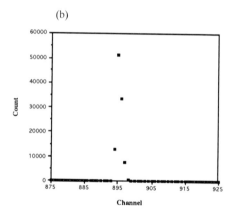

(c)

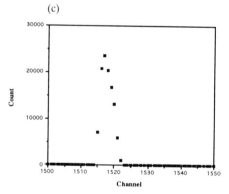

Figure 2.10 Histogram plots for the tungsten–halogen lamp; thermal light source: (a) 3 V input; (b) 6 V input; (c) 8 V input.

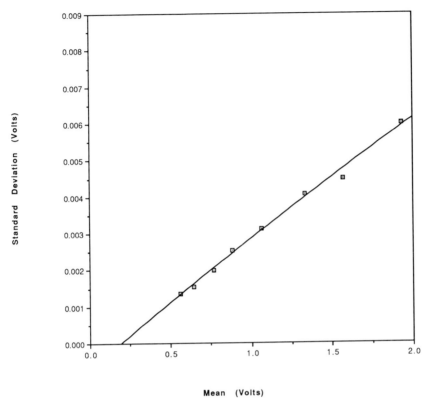

Figure 2.11 Standard deviation versus mean output voltage for the helium–neon laser.

binary, they have two stable states, 'on' or 'off'. Additional intermediate levels can be created by area modulation. Active devices are typically composed of a large number of individual pixels each of which can be turned 'on' or 'off'. Area modulation involves using an area containing a number of pixels. By changing the ratio of pixels which are 'on' to those which are 'off', intermediate gray levels can be created. Four active SLMs were used in the experiments. Each will be discussed later in this section. One passive SLM was used, the neutral density filters (Kodak Filters for Scientific and Technical Uses, 1976). These filters comprise colloidal carbon suspended in a gelatin. The filters used were the Kodak Wratten neutral density filters which are neutral between 700 and 400 nm and had a manufacturing tolerance of ± 5%.

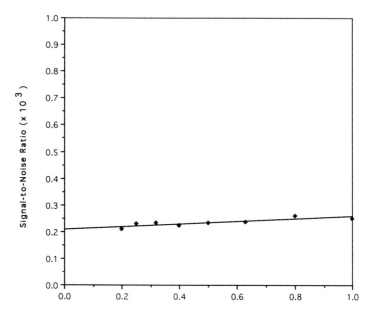

Figure 2.12 SNR for the helium–neon laser.

p-i-n photodiode detectors

The p-i-n photodiode used was a Hamamatsu model S1188-02. This device had a 0.8 mm diameter with a 0.5 mm^2 effective area. It was sensitive to wavelengths between 400 and 1060 nm with the peak sensitivity wavelength at 900 nm. The p-i-n photodiode was chosen for use in the experimental measurements because it had a lower dark current than other photodiodes. For this device, the dark current was 2 nA at 25 °C (Photodiodes, 1987).

Multichannel analyzer

The multichannel analyzer used was a Canberra model 35 Plus (Series 35 Plus Multichannel Analyzer Operator's Manual Version 2, 1985). The multichannel analyzer was operated in the sampled voltage analysis (SVA) mode for generation of the histograms. In this mode, an external gate pulse was applied to the device which sampled the incoming voltage coincident with the trailing edge of the gate pulse.

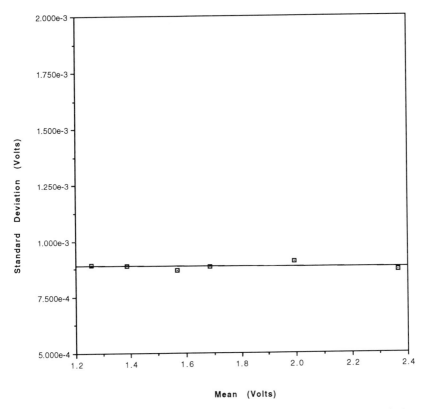

Figure 2.13 Standard deviation versus mean output voltage for the laser diode.

The device could sample voltages between 0 and 10 V and store the sample in one of 2048 memory locations (channels). The multichannel analyzer could also be set to operate as if it had 4096 or 8192 channels. In these cases, the FSRs for sampling were 0–5 V and 0–2.5 V, respectively. The device was used in the latter setting which provided a sensitivity of 1.22 mV separation between channels. The external gating pulse was a 5 Hz square wave with a pulse amplitude of 7 V. The time window in which the device would sample the incoming voltage could be set by the operator. This window was set at 20 s to provide a large number of counts (approximately 100 000 samples) to ensure accurate statistics. A PC computer was used to control all operating parameters of the multichannel analyzer. Canberra provided the control software which was written in BASIC.

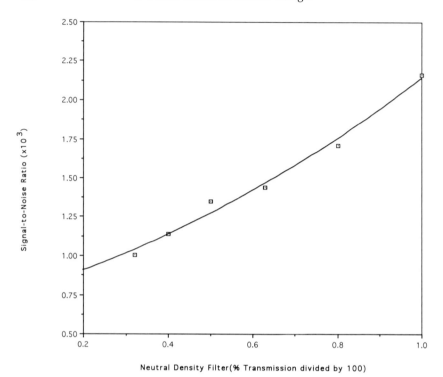

Neutral Density Filter(% Transmission divided by 100)

Figure 2.14 SNR for the laser diode.

2.5.2 Liquid crystal light valve

The liquid crystal light valve (LCLV) tested was a Hughes model H-4050. The LCLV was an optically addressed liquid crystal device (Bleha *et al.*, 1978). The device was composed of a CdS photoconductor, a CdTe light-absorbing layer, a dielectric mirror, and a biphenyl liquid crystal layer sandwiched between indium-tin-oxide transparent electrodes deposited onto optical quality glass flats. Information is imposed on the incident light beam called the 'write' beam.

The CdS photoconductor acts as a light-variable resistor so that a variable applied field is dropped across the CdS and hence appears across the liquid crystals. This, in turn, causes the liquid crystals to rotate in those regions corresponding to where the light is incident on the photoconductor. It does not affect those areas where the light is not incident. Another light beam, the 'read' beam, is incident on the opposite face of the device. The light from the read beam passes through the liquid crystal and is reflected back through the liquid crystal material by a dielectric mirror that separates the liquid crystal material from the photoconductor material. In traversing the liquid crystal, the polarization of the read beam is rotated 90° in those areas where no voltage is applied and is not rotated in those areas where voltage is applied. As a polarizer is placed between the light source and the LCLV, only the unrotated light leaving the LCLV will pass through without being blocked. This creates the 'on' and 'off' states. A beamsplitter was used to divert the read beam away from the light source and toward the detector. The LCLV required a sinusoidal bias voltage to operate, which was set at 2.47 V RMS at 2 KHz for these experiments. All measurements were made using the tungsten–halogen lamp with a polarizer–analyzer pair.

Figure 2.15 contains plots of the standard deviation versus the mean output voltage for the LCLV in both the 'on' and 'off' states. Both

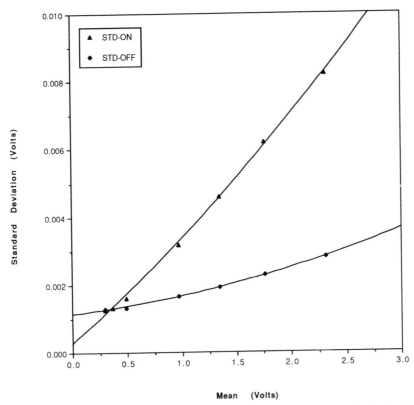

Figure 2.15 LCLV standard deviation versus mean output voltage for 'on' and 'off' states.

states show the standard deviation to be signal-dependent. The 'on' state shows a stronger signal-dependence than the 'off' state. Figure 2.16 plots the SNR versus the input voltage (read beam) for both the 'on' and 'off' states. Again, the standard deviation in the 'on' state can be seen to be increasing such that the SNR decreases with increasing light intensity of the read beam on the SLM. In Figure 2.17, the histograms corresponding to three input voltages are plotted. As the light intensity increases one notes that a second peak begins to form. This twin peak histogram reflects the superposition of the sinusoidal modulation of a noisy signal with a stronger noisy unmodulated signal. The only source of the sinusoidal modulation is the AC biasing signal used by the LCLV. This signal is weakly modulating the light intensity of the read beam. Figure 2.18 shows the accuracy bounds for the LCLV device. Note that in Figure 2.18, as well as subsequent accuracy

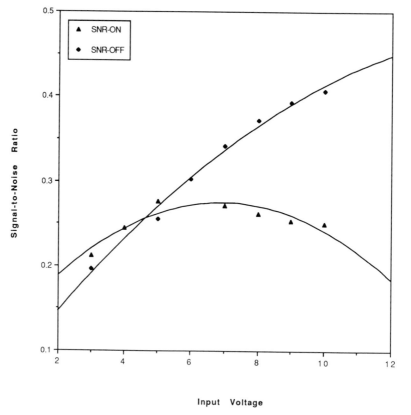

Figure 2.16 SNR versus input voltage for LCLV.

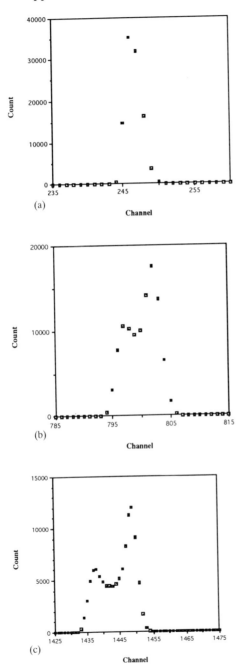

Figure 2.17 Histogram plots for the LCLV: (a) 3 V input; (b) 7 V input; (c) 9 V input.

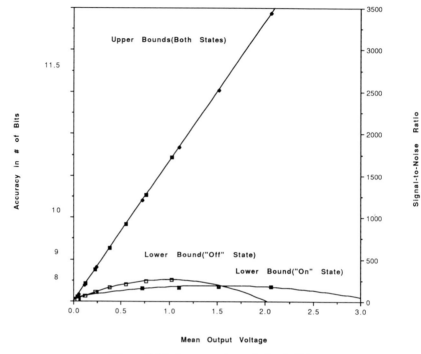

Figure 2.18 Accuracy bounds for LCLV.

bound figures, accuracy in number of bits is shown on the right vertical axis in a nonlinear scale while the SNR is on the left vertical axis in a linear scale. In Figure 2.18 the temporal modulation has greatly increased the standard deviation of the read beam. The overall accuracy – that is the ability to relate a numerical value to a light intensity level – has been greatly decreased. For this reason the usable dynamic range of this device in an OLAP is less than the physical dynamic range achievable with the device by several bits.

2.5.3 Liquid crystal television

The liquid crystal television (LCTV) was a Radio Shack Pocketvision 5 (Boreman & Raudenbush, 1988; Liu, Davis & Lilly, 1985; Liu & Choa, 1989). This model had 146 × 120 pixels where each pixel is a 90° twisted nematic liquid crystal cell that is optically transmissive. All cells are surrounded by a wire grid structure through which current can be applied to generate an electric field. When light passes through a liquid crystal cell with no electric field present, its polarization is rotated by 90°, while with an electric field applied no polarization rotation occurs. For linearly polarized light incident on the device, a polarizer can be placed after the device to block one polarization orientation and not the other, creating 'on' or 'off' states. As the device is primarily used as a television, a National Television Standards Committee (NTSC) raster scan format is used at a 30 Hz rate to update the pixels. After the field is applied the crystals begin to rotate back to the nonfield state until the field is applied again. Figure 2.19 is a plot of the standard deviation versus mean output voltage for both the 'on' and 'off' states. The standard deviations for both states are strongly signal-dependent. Figure 2.20 is a plot of the SNR versus the input voltage for both the 'on' and 'off' states. For both states the SNR decreases with increasing intensity. The 'off' state initially has an increasing SNR

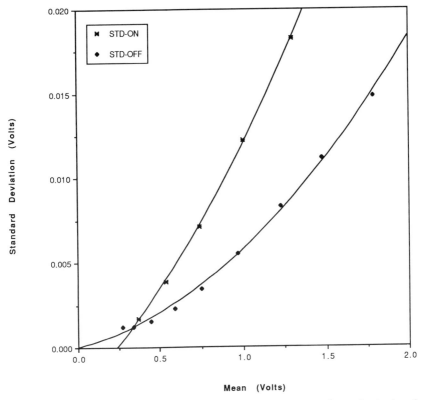

Figure 2.19 LCTV standard deviation versus mean output voltage for 'on' and 'off' states.

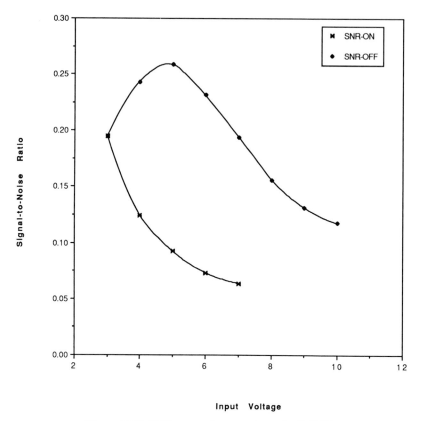

Figure 2.20 SNR versus input voltage for LCTV.

at low intensities as noise terms other than the standard deviation of the light incident on the detector dominate in the standard deviation of the output voltage.

Figure 2.21 shows the output voltage histograms for the LCTV. It is important to note that the histogram of a constant slope ramp wave-form has a characteristic flat (square wave) shape. This shape is evident in the histograms at 7 and 10 V and is due to the NTSC raster scan used to refresh the pixels. The transmissivity is at a maximum when refreshed. It maintains this value for a time and then begins to decay at a constant rate. At some point the decay rate changes as two slopes can be seen from the histograms. The pixel is then refreshed again and the process is repeated. This process temporally modulates the light intensity passing through the SLM and is the primary source

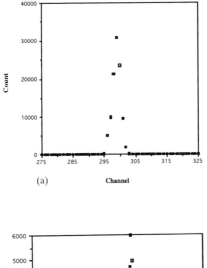

(a)

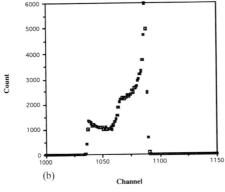

(b)

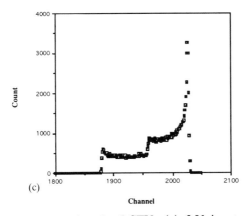

(c)

Figure 2.21 Histogram plots for the LCTV: (a) 3 V input; (b) 7 V input; (c) 10 V input.

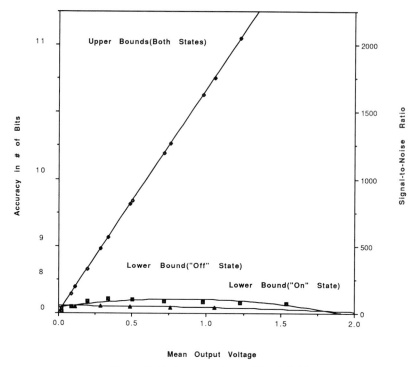

Figure 2.22 Accuracy bound for LCTV.

of noise in the device. The accuracy bounds are shown in Figure 2.22 and show that the achievable accuracies are very low due to this temporal modulation effect.

2.5.4 Ferro-electric liquid crystals

The ferro-electric liquid crystal (FLC) spatial light modulator studied was manufactured by STC Technology, Ltd, and loaned to us by DisplayTech, Inc. (Pagano-Stauffer, 1987; Instruction Manual 128 × 128 Ferroelectric Liquid Crystal Spatial Light Modulator, 1989; Handschy, Johnson & Cathey, 1987; Johnson, 1987). The model used had 128 × 128 pixels. The liquid crystal cell consisted of a 1.7 μm layer of ferroelectric chiral smectic C liquid crystal between two sheets of optically flat glass. The inner surfaces of the glass were coated with a transparent electrically-conducting indium-tin-oxide film to allow control voltages to be applied across a pixel.

When light passes through the liquid crystal while a voltage is applied across the material, its polarization is rotated by 45°. If the voltage is applied having the opposite polarity, its polarization is rotated by 45° in the opposite direction. When linearly polarized light passes through the device, the analyzer can be rotated such that light polarization in one plane passes unattenuated while that in the other plane is greatly attenuated. In this way, 'on' and 'off' states can be created. Pixels were turned 'on' and 'off' using an IBM PC computer acting as a controller, with software provided by DisplayTech. The SLM had its own power supply which provided an AC biasing signal, the frequency of which was controlled by a potentiometer on the device. The standard deviation versus mean output voltage for both the 'on' and 'off' states are shown in Figure 2.23. The standard

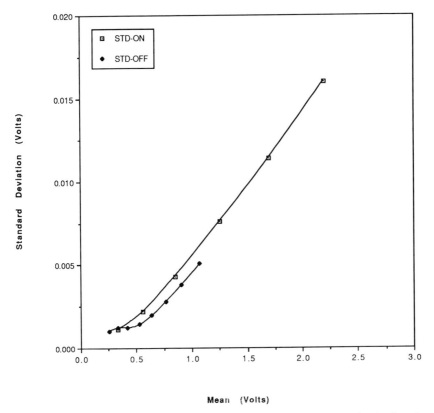

Figure 2.23 FLC standard deviation versus mean output voltage for 'on' and 'off' states.

deviations for both states were again strongly signal-dependent. Figure 2.24 is a plot of the SNR versus the input voltage for both the 'on' and 'off' states. The hump-shaped curve for the 'off' state is due to the thermal noise process dominating at the lower intensity levels and the standard deviation of the incident light on the detector dominating at the higher intensity levels. The histograms are shown in Figure 2.25. Again a twin peak histogram can be seen. In this case the constant intensity predominates and the sinusoidal modulation has a much smaller effect than for the LCLV but still causes the standard deviation to increase with increasing light intensity. The strong signal-dependency of the light from the FLC SLM results in poor accuracy performance as can be seen in Figure 2.26.

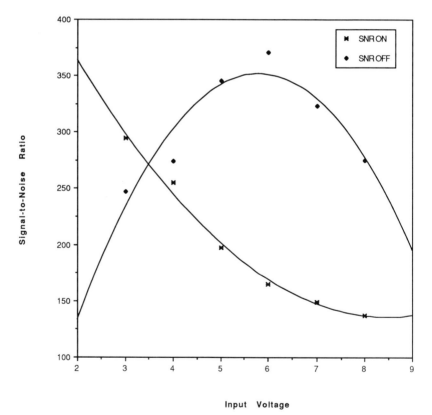

Figure 2.24 SNR versus input voltage for the FLC SLM.

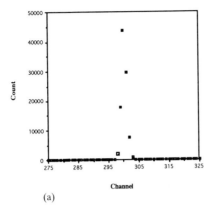

(a)

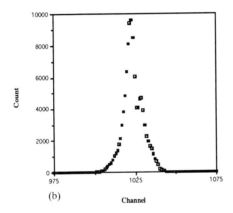

(b)

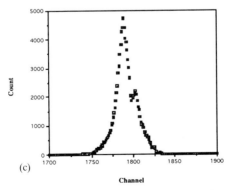

(c)

Figure 2.25 Histogram plots for the FLC SLM: (a) 3 V input; (b) 6 V input; (c) 8 V input.

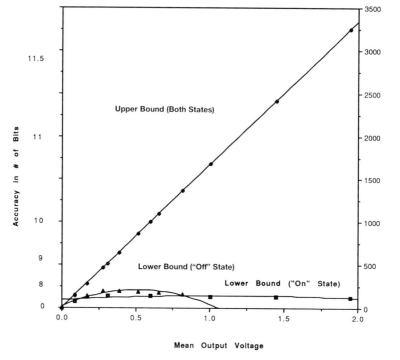

Figure 2.26 Accuracy bounds for the FLC SLM.

2.5.5 Magneto-optic SLMs

The magneto-optic spatial light modulator (MOSLM) used was manufactured by Semetex Corporation (Specifications and Performance
Updates, 1988; Ross, Psaltis & Anderson, 1983; Davis, Gamlieli &
Bach, 1988). We used the 128×128 pixel model. The active material
was an iron–garnet crystal that is optically transparent. The material
has two magnetic states that rotate the polarization of the light
clockwise or counterclockwise by an equal amount. This is accomplished by using Faraday rotation. For the device used in the experiments, the Faraday rotation was $14°$ at 632 nm and $32°$ at 546 nm.
When linearly polarized light is incident on the front of the device an
analyzer can be rotated so that the light from one magnetic state passes
through unaffected while light from the other magnetic state is greatly
attenuated. This results in the 'on' and 'off' states. A percentage of the
incident light is absorbed as it passes through the material. For our
device, this was 32% at 632 nm and 61% at 546 nm. Which magnetic

state exists in a particular pixel is controlled electrically and can be switched at 100 ns using software provided by Semetex. Figure 2.27 shows plots of the standard deviation versus mean of the output voltage for both the 'on' and 'off' states. The SNR versus the input voltage for the MOSLM is shown in Figure 2.28. While the standard deviations for both states were again signal-dependent, the SNRs increased with increasing intensity and the histogram plots in Figure 2.29 show no sign of temporal modulation effect. This results in a better accuracy performance than the other active SLMs as can be seen in Figure 2.30. The principal difficulties found with the MOSLM were light leakage between the pixels and high light attenuation. The first problem had been corrected by Semetex and the latter problem can be reduced by the use of laser light at wavelengths in the red and green bands that optimize the MOSLM CR.

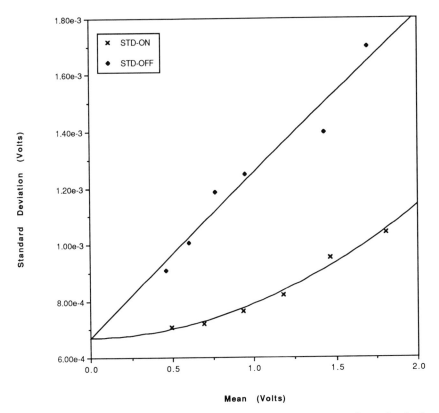

Figure 2.27 MOSLM standard deviation versus mean output voltage for 'on' and 'off' states.

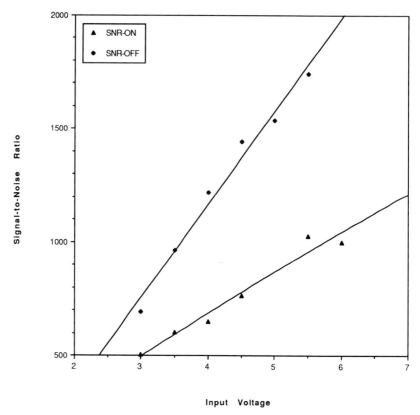

Figure 2.28 SNR versus input voltage for the MOSLM.

2.6 Conclusions

In an OLAP the light intensity level at each point in the processor is related to a numerical value. Uncertainty due to fluctuations in the light intensity and in the detection process limits the accuracy of the results. The light intensity can be modeled as a random variable in which the mean value is related to the desired signal level and the standard deviation is related to the noise.

In linear algebra operations, in particular, matrix–vector and matrix–matrix multiplication, two operations are required: addition and multiplication. By analyzing these operations using second order statistics, it has been shown that the multiplication operation generates signal-dependent noise that has a stronger signal dependency than that generated in the detection process. Furthermore, as the light intensity

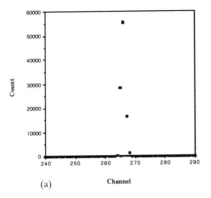

(a)

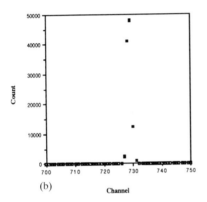

(b)

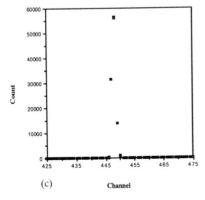

(c)

Figure 2.29 Histogram plots for the MOSLM: (a) 5 V input; (b) 12 V input; (c) 8 V input.

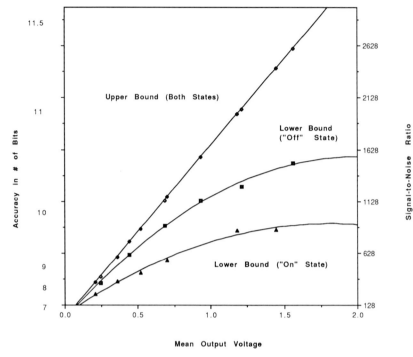

Mean Output Voltage

Figure 2.30 Upper and lower accuracy bounds for the MOSLM.

increases it becomes the dominant noise term. Using the statistical
results, design criteria such as SNR, dynamic range, and accuracy
bounds can be formulated.

By experimentally measuring the output voltage of a single OLAP
channel for a particular set of components (light source, SLM, and
detector), a histogram can be generated and the mean and standard
deviation can be determined as a function of light intensity at the
detector. Four active SLMs were measured using this method and the
results compared. In three of the four SLMs tested, we found that in
an effort to improve the CR either an AC bias signal or a raster scan
signal were used. In all three cases where the output light intensity was
being temporally modulated by this signal, the accuracy and the usable
dynamic range of the systems were severely limited. Only the MOSLM
which used no bias or scanning signal showed a reasonable accuracy
over a range of light intensities. Many of the difficulties with these
SLMs stem from the fact that optical computing was not the primary
application for which they were designed. Rather, they were designed

for image processing and projection television applications where CR is very important and the temporal modulation effects discussed here are not visible to the user. As optical computing becomes the principal application for SLM design, noise properties that affect accuracy must be minimized in the design process.

Successful design of a practical optical processor requires that the noise properties of each of the components as well as their interactions as a system be taken into account. By using experimentally measured values for the mean and standard deviation of the components, and the system analysis described above, design trade-off studies can be performed that will determine the optimum processor architectures and the selection of components to meet the design criteria.

2.7 Acknowledgements

The authors wish to acknowledge the support, at Texas Tech University, of this research by the Office of Naval Research under the SDIO/IST program.

References

Athale, R. A., Optical Matrix Processors. In *Optics and Hybrid Computing*, Harold H. Szu, ed., *Proc. SPIE, Vol. 634*, p. 96, SPIE, Bellingham, WA (1986).

Athale, R. A., and Collins, W. C., Optical Matrix–Matrix Multiplier based on outer product decomposition, *Appl. Opt.*, **21**, No. 12, 2089 (1982).

Batsell, S. G., *Accuracy Limitations in Optical Linear Algebra Processors*, Ph.D. Dissertation, Texas Tech University, Lubbock, May (1990).

Batsell, S. G., Jong, T. L., Walkup, J. E., and Krile, T. F., Comparison of Noise Due to Passive and Active Spatial Light Modulators in Optical Matrix–Vector Multipliers. In *OSA Annual Meeting*, Technical Digest Series, Vol. 11, Optical Society of America, Washington, DC, p. 72 (1989a).

Batsell, S. G., Jong, T. L., Walkup, J. F., and Krile, T. F., Fundamental Noise Limitations in Optical Linear Algebra Processors. OSA Annual Meeting, 1989 Technical Digest Series, Vol. 18, Optical Society of America, Washington, DC, p. 118 (1989b).

Batsell, S. G., Jong, T. L., Walkup, J. F., and Krile, T. F., Noise Limitations in Optical Linear Algebra Processors, *Appl. Opt.*, **29**, 2084–2090 (1990).

Bleha, W. P., Lipton, L. T., Wiener-Avnear, E., Grinberg, J., Reif, P. G., Casasent, D., Brown, H. B., and Markevitch, B. V., Application of the Liquid Crystal Light Valve to Real-Time Data Processing, *Opt. Eng.*, **17**, No. 4, 371 (1978).

Boreman, G. D., and Raudenbush, E. R., Modulation Depth Characteristics of a Liquid Crystal Television Spatial Light Modulator, *Appl. Opt.*, **27**, No. 14, 2940 (1988).

Casasent, D., and Ghosh, A., Optical linear algebra processors: noise and error-source modeling, *Opt. Letters*, **10**, 252 (1985).

Caulfield, H. J., and Rhodes, W. T., Optical Algebraic Processing Architectures and Algorithms, John A. Neff, ed., *Proc. SPIE, Vol. 456*, p. 2, SPIE, Bellingham, WA (1984).

Davis, J. A., Gamlieli, J., and Bach, G. W., Optical Transmission and Contrast Ratio Studies of the Magnetooptic Spatial Light Modulator, *Appl. Opt.*, **27**, No. 24, 5194 (1988).

Dias, A. R., Incoherent Matrix–Vector Multiplication for High-Speed Data Processing, Tech. Report No. L722-4, Stanford University (1980).

Gersho, A., Principles of Quantization, *IEEE Trans. Circuits and Systems*, **CAS-25**, No. 7, p. 427–435 (1978).

Goodman, J. W., *Statistical Optics*, John Wiley, New York (1985).

Goodman, J. W., and Woody, L. M., Methods for Performing Complex-Valued Linear Operations on Complex-Valued Data Using Incoherent Light, *Appl. Opt.*, **16**, No. 10, 2611 (1977).

Goodman, L. M., Dias, A. R., and Woody, L. M., Fully Parallel, High Speed Incoherent Optical Method for Performing Discrete Fourier Transforms, *Opt. Letters*, **2**, No. 1, 1 (1978).

Goodman, J. W., Dias, A. R., Woody, L. M., and Erickson, J., Some New Methods for Processing Electronic Image Data using Incoherent Light, *Proc. ICO-11*, Madrid, Spain, 139 (1978).

Handschy, M. A., Johnson, K. M., and Cathey, W. T., Polarization-based Optical Parallel Logic Gate Utilizing Ferroelectric Liquid Crystals, *Opt. Letters*, **12**, No. 8, 611 (1987).

Instruction Manual, 128 × 128 Ferroelectric Liquid Crystal Spatial Light Modulator, STC Technology Ltd, Essex, England (1989).

Johnson, K. M., Optical Computing and Image Processing with Ferroelectric Liquid Crystals, *Opt. Eng.*, **26**, No. 5, 385 (1987).

Jong, T. L., Walkup, J. F., and Krile, T. F., Noise Effects in Optical Linear Algebra Processors, *J. Opt. Soc. Am. A.*, **3**, 24 (1986).

Kodak Filters for Scientific and Technical Uses, Eastman Kodak Company, Rochester, NY, (1976).

Laser Diode User's Manual, Sharp Electronics Corporation, Mahwah, NJ (1986).

Liu, H. K., and Choa, T. H., Liquid Crystal Spatial Light Modulators, *Applied Optics*, **28**, No. 22, 4772 (1989).

Liu, H. K., Davis, J. A., and Lilly, R. A., Optical Data-Processing Properties of a Liquid Crystal Television Spatial Light Modulator, *Opt. Letters*, **10**, No. 12, 635 (1985).

McAulay, A. D., Spatial-Light-Modulator Interconnected Computers, *Computer, Vol. 20*, No. 10, 45 (1987).

Pagano-Stauffer, L. A., *Optical Logic Gates Using Ferroelectric Liquid Crystals*, Master's Thesis, University of Colorado, (1987).

Panasonic Data Sheet, *Digi-Key Corporation Catalog*, Digi-Corporation, Thief River, MN (1988).

Perle, C. J., and Casasent, D. P., Effects of Error Sources on the Parallelism of an Optical Matrix–Vector, *Appl. Opt.*, **29**, 2544 (1990).

Photodiodes, Hamamatsu Photonics Corporation, Bridgewater, NJ (1987).

Ross, W. E. Psaltis, D., and Anderson, R. A., Two-Dimensional Magneto-Optic Spatial Light Modulator for Signal Processing, *Opt. Eng.*, **22**, No. 4, 485 (1983).

Seifert, E., and Nauda, A., *Enhancing the Dynamic Range of Analog-to-Digital Converters by Reducing Excess Noise*, IEEE Pacific Rim Conf. on Comm., Comp., and Signal Processing, p. 574–6 (1989).

Series 35 Plus Multichannel Analyzer Operator's Manual Version 2, Canberra Industries, Meriden, CT (1985).

Specifications and Performance Updates, Semetex Corporation, Torrance, CA (1988).

Spenser, N., *Comparison of State-of-the-Art Analog-to-Digital Convertors*, MIT Lincoln Laboratory Report AST-4, 4 March 1988.

Taylor, B. K., and Casasent, D. P., Error-Source Effects in a High Accuracy Optical Finite-Element Processor, *Appl. Opt.*, **25**, 966, (1986).

Van Trees, H. L., *Detection, Estimation, and Modulation Theory*, Part I, John Wiley, New York (1968).

Whalen, A. D., *Detection of Signals in Noise*, Academic Press, New York (1971).

Zuch, E. L., ed., *Data Acquisition and Conversion Handbook*, Datel-Intersil, Mansfield, Mass. (1979).

3

Effects of diffraction, scatter, and design on the performance of optical information processors

R. B. BROWN and J. N. LEE

Naval Research Laboratory, Optical Sciences Division

3.1 Introduction

Optical processors hold tremendous potential for processing many information channels in parallel, each at very high bandwidth, and in small volumes with low power consumption (VanderLugt, 1992). Major classes of operations that optics can perform include integral transformations such as Fourier transforms and correlations, and matrix operations, e.g. vector–matrix multiplication (Lee, 1987). Many, but not all, of these are made possible by the remarkable property that a 'Fourier transform lens' generates, at the back focal plane, the two-dimensional (2-D) Fourier transform of phase and amplitude information input at the front focal plane.

Initial feasibility demonstrations of optical signal processors are usually performed in the laboratory using existing components, but these often do not come close to fully exploiting the potential of the optical processor, and often disregard practical implementation issues such as: how much phase distortion is permissible in a critical lens, or what will be the effect of vibration or temperature changes in the field, and is it feasible to have a given dynamic range and information capacity in the same processor? Therefore the aim of most subsequent optical hardware development is to answer these questions and overcome any deficiencies.

Much effort has gone into the development of the critical active optical devices, such as infrared laser diodes, spatial light modulators (SLMs) and photodetector arrays, and new device concepts and improvements on existing devices continue to occur (Lee & Vander-Lugt, 1989). High-performance optical components, such as high-dynamic-range photodetector arrays (Anderson, Guenther, Hynecek, Keyes & VanderLugt, 1988) and high-power semiconductor laser diodes (Scifres, Lindstrom, Burnham, Streifer, & Paoli, 1983), have been used in conjunction with SLMs to demonstrate system capabilities.

The design, or integration process, for optical processors turns out to be extremely crucial to achieving improved levels of performance and maintaining this performance over time in real-world environments where temperature and vibration are not controlled. When active optical devices are integrated with passive optical components (mirrors, lenses, prisms, etc.) and joined by mechanical structures into an optical system, actual performance will be determined by how well the optical design integrates the processing requirements, the active and passive optical components, and the environmental disturbances.

This chapter will deal with issues which limit the basic processing capability of an optical processor and which must be considered when designing for extremely high performance in the real world. These issues include: (1) *optical design procedures and software* and their ability to handle standard design concepts such as geometrical distortion, as well as quantities peculiar to some optical processors, such as absolute phase accuracy, and the proper modeling or calculation of theoretical and predicted performance; (2) *fabrication limitations* such as surface and assembly accuracy, glass homogeneity, etc., affecting how well the theoretical and predicted performance agree; (3) *physical optics effects*, e.g., diffraction and stray light such as from optical scatter, which affects the trade-off between information density and dynamic range, as well as the maximum theoretical performance. Stray light can now be modeled using certain optical software and therefore can be included in the design process. Very important considerations that will not be addressed include the efficiency of management of the optical processor and pre- and post-(electronic) processing.

The type of optical processor that has undergone the most development is the acousto-optic (AO) type and is the type referred to in most examples in this chapter. Although AO cells are only one-dimensional (1-D) devices, they have enabled the exploitation of some of the bandwidth and parallel-channel capacity of optics. The most common AO devices are bulk Bragg cells and surface acoustic wave (SAW) devices. Other processor types include those using fixed 2-D array SLMs and laser diode/photodiode arrays. AO processors exemplify many of the major aspects of optical processing: all the major operations that optics can do have been performed using AO, e.g., Fourier transformations, correlation, and matrix operations as high-performance building blocks for a large variety of algorithms (Lee & Vander-Lugt, 1989), including 2-D processing algorithms. AO processors are distinguished mainly by the fact that there exist large-aperture, high-bandwidth AO cells known generally as Bragg cells (Chang & Lee, 1983) which can temporally and/or spatially modulate a laser beam. These cells employ the photoelastic interaction between light and an information-carrying ultrasonic wave in a crystalline or glass medium amplitude and phase modulate the light passing through that medium. Demonstrations of AO technology for high-performance include the heterodyne doubly-diffracting broadband correlator (Griffin & Lee, 1994) and the high performance AO rf channelized receiver (Anderson, Webb, Spezio, & Lee, 1991). Design issues for these two optical processor examples will be treated in the following sections.

Since the desired optical design is usually the least expensive one which meets the performance requirements, it is very useful to have an accurate computer model to give meaningful performance estimates during the design process. This is primarily due to the not always obvious interaction of various optical performance factors, which for some processors may include size, weight, temperature stability, and phase jitter, in addition to standard optical performance. While it is obvious that better optical performance can always be obtained by using higher-quality optical components and putting more effort into design, if it is not known how this effort benefits the processor performance the effort will probably be wasted. Quantitative estimates of performance are needed to ensure that the design effort is not wasted by poor performance in the finished system, or by more effort than is actually needed to reach the performance goals. Hence, a parameter such as 'dynamic range' must include consideration of channel isolation or selectivity, specification of linearity, and the required small-signal sensitivity. These restrictions can result in a trade-off against the perceived high channel densities in optical systems.

3.2 Common processor performance parameters

Time aperture

This is the maximum length of data that can be processed in parallel. In an AO processor, it is the length of the acoustic beams(s) illuminated by the laser beam, divided by the acoustic velocity of the cell(s), usually expressed in microseconds. It is not always necessary to wait for the aperture to fill before the data can be processed; for example, the AO correlator used as an example in this chapter can continuously generate the correlation function of two continuous sets of data in a 50 μs time aperture. Time aperture is limited by the ability to grow, fabricate, and mount large crystals with sufficiently good optical quality, and by acoustic diffraction and absorption, which places limits on the length and uniformity of the acoustic beam. Time apertures of existing AO cells can range from the optical diffraction limit (point modulators) up to about 100 μs for shear tellurium dioxide (TeO$_2$). In processors using 2-D SLMs in place of AO cells, it is the number of pixels and the frame or addressing rate of the device that determines the processing capacity. In AO processors, the bandwidth of the AO cells is the other important factor in determining processing capacity. Also, since AO processors are generally limited to an octave bandwidth, higher-frequency devices have potentially greater bandwidth.

Time-bandwidth product (TBWP) and processing capacity

The parallel-processing capacity of an AO processor for processing 1-D information is ultimately limited by the product of the temporal bandwidth of the AO cells (in cycles per second) and the time aperture (in seconds), hence TBWP is the (unitless) number of 'spots' or parallel optical channels. Usually, however, the effective TBWP is less than this because it is limited by the optical design of the processor or other physical factors to be discussed later. In this case the TBWP can be shown to be directly related to the optical diffraction limit of the illuminated aperture, i.e. how closely one can pack discrete optical channels (Lee & VanderLugt, 1989).

The TBWP of existing Bragg cells can be several thousands. At one extreme are cells made of material with low acoustic velocity, v, and high acoustic attenuation. A prime example is shear mode TeO$_2$ ($v = 617$ m/s) which can have time apertures up to 100 μs if operated below 100 MHz. But bandwidths are limited to approximately 40 MHz due to the low operating frequencies and octave bandwidth limits. Thus, TBWP for this material could be as high as 4000 if the expense and the acoustic nonuniformity of such a long cell can be tolerated. New mercurous halide materials (Goutzoulis, Gottlieb, and Singh, 1992), such as mercurous chloride (Hg$_2$Cl$_2$), with $v = 347$ m/s, could result in cells with even higher TBWPs, when optical quality and fabrication techniques are fully developed.

At the other extreme, cells made of material with a much lower acoustic loss coefficient can be operated at higher frequencies, limited by the frequency-squared dependence of absorption. However, the acoustic velocity is usually much higher and the fall-off in acoustic intensity with distance now severely limits the usable time aperture. A good example is longitudinal mode lithium niobate (LN), which can be operated at 2 GHz center frequency with 1 GHz bandwidth, but the time aperture is limited to about 1.0 μs, giving a TBWP of 1000. Another high-frequency material, with a figure of merit, M_2, six times larger than LN and therefore of greater potential, is gallium phosphide (GaP), but its acoustic loss is highly dependent on the quality and source of the material, which have been highly variable. Cells made of these materials have been available from sources in the US and elsewhere for several years.

Channel dynamic range

A large TBWP equates to a large number of parallel information channels that can be processed simultaneously. Optics provides the ability to process and interconnect these large numbers of high-bandwidth channels. Practical realization of this capability depends on having low optical signal crosstalk between channels so that channel performance, i.e., channel dynamic range, is not seriously restricted by crosstalk. However, crosstalk, in various forms which will be discussed, is usually the ultimate limitation on channel dynamic range, but not necessarily processor dynamic range. In space, or in a linear medium at low optical intensities, closely associated or crossing optical beams do not interfere with or corrupt information on each other, and this has led to the widespread belief that optics should be employed for low crosstalk. However, it is only electromagnetic crosstalk (or electromagnetic interference, EMI) that is eliminated by using optical channels instead of electrical channels and does not mean that there are no crosstalk mechanisms in optical systems.

Selectivity

High dynamic range and a large number of channels equates to high selectivity, or the ability to process accurately a weak signal adjacent to a strong signal. High selectivity is clearly important in a 1-D channelizer, or rf spectrum analyzer, if it is necessary to process or detect weak signals which are adjacent to strong signals.

Channel crosstalk and diffraction

In closely packed parallel optical channels, a fundamental limitation is due to the overlap of the diffraction 'tails' of the channels at a focal plane where a photodetector array or other device is located. Even 'diffraction-limited' channels may contribute more than the desired crosstalk to adjacent channels, or beyond, if the shape of the diffraction tail is not controlled or taken into account in the design process. The magnitude of diffraction crosstalk depends on details of the optical system, since the diffraction patterns are due to truncations of the light beam by edges and opaque barriers. Low crosstalk equates to high dynamic range, but is not the only factor which limits it. Also diffraction is not the only cause of crosstalk.

Channel crosstalk and scattered light

Scattered light, which is stray light caused by imperfections such as surface roughness, glass index inhomogeneities, inclusions, subsurface damage, coatings, or dirt, is a second factor which can contribute to crosstalk and limiting the dynamic range. Traditionally, scattered light has not been dealt with in optical computer aided design (CAD) packages used by lens and system designers, but by specialized software for stray light analysis only. This software has been expensive and requires additional training to use it. It has been used to check the stray light performance of high-cost systems such as space telescopes and solar coronagraphs where the time and expense are clearly justified. More recently, new optical design software, together with new hardware for characterizing the small-angle scattering of components and materials, have made it possible to examine the effects of stray light early in the design process more easily. These will be described later in the chapter.

Channel crosstalk and stray light

Another type of stray light which can limit dynamic range is due to unwanted residual surface reflections from every surface in the processor. The most troublesome in any imaging system, because they are forward propagating toward the image, are 'first order' ghost images due to all possible combinations of two reflections. These are important in processors too because some may have a focus near a detector plane, giving stray signals strong enough to limit dynamic range. In a processor with a sensitive laser source, single reflection, or 'zero order', ghost images are also important because they may direct light back into the laser, causing instability and noise. First order ghost image analysis is included in some standard optical CAD packages used by designers. Zero order is not. Section 3.3 lists some of the standard and special-purpose CAD packages available and some of their more important capabilities.

Processor dynamic range

In a single (Fourier) transform processor such as the rf spectrum analyzer, the channel dynamic range is also the effective dynamic range of the processor. In multiple transform processors there may be multiple factors which combine to give the effective dynamic range. Also it should be mentioned that dynamic range can be limited by other than optical factors: unwanted intermodulation distortion (IMD) signals generated by nonlinearity in AO materials, or crosstalk caused by vibration-induced changes in the optical path. These IMD signals are analogous to those in electronic amplifiers.

3.3 CAD tools for design and analysis

3.3.1 Standard optical design/analysis packages

There exist both standard and special-purpose optical CAD and analysis packages. Standard CAD packages cover a wide range of capability, from very inexpensive ray trace packages which can handle a limited-size optical system with many of the most basic capabilities listed in Table 3.1, to very comprehensive packages with most of the features listed, plus many others. In the table more universal or basic features are generally listed first. The most capable packages present very few limitations to the lens designer who is interested in laying out, optimizing, analyzing, and preparing fabrication drawings and data for the most complicated lens. If the package does not have optimization, sometimes called 'automatic design', it is not very useful as a lens design tool but will still be useful for analysis of existing designs and some 'manual' operations. Optical CAD packages have evolved primarily to design lenses for incoherent imaging, since that is still the primary application of most lenses. Consequently, some features useful for coherent design and analysis may be difficult to use or not available in some packages. Lenses for incoherent imaging purposes do not require careful control of phase between the input and output plane; if the optical path variation, or optical path difference (OPD) through

Table 3.1 Capabilities of optical design/analysis packages.[a]

System types	Single axis, multiple axis symmetrical, nonsymmetrical, global coordinates.
Surface type	Plano, spherical, cylindrical, aspherical, tilt, decenter, diffracting, holographic. Scattering, optical.[b] Scattering, nonoptical.[b]
Surface coatings	None, or all reflective or transmissive, multi-layer. Specify reflection, transmission, and absorption.[b]
Glass data	Some glasses, complete catalog, multiple catalogs, IR materials, gradient index glasses. Birefringent (uniaxial) materials.[b]
Ray trace	Real, real and paraxial, single. Gaussian beam propagation, exit pupil wavefront map, gradient index trace. Multiple Gaussian beam trace to permit general diffraction solutions.[b]
Ray-based analysis	Third order aberration plots, spot diagrams, encircled energy, geometrical MTF & point spread function, first order ghost images, polarization. Ghost images, any order.[b]
Diffraction-based analysis	Single Gaussian beam propagation, diffraction PSF (diffraction solution at focal plane), with detector integration, encircled energy, diffraction MTF, partial coherence. Gaussian beam decomposition/synthesis (diffraction solution at any location).[b]
Optimization, or 'automatic design'	None, local ray-based, local diffraction-based, global (not limited to local optimum).
Environmental	Temperature, pressure, vibration.[c]
Fabrication	Automatic tolerancing/sensitivity analysis, system and component drawings, alignment aids, weight/cost analysis.

[a]Roughly in order of increasing sophistication. [b]Special analysis packages. [c]Mechanical analysis performed externally.

the lens, for each object/image point is $\lambda/4$ or less, the image will be very good and it can be called 'diffraction-limited'. However, even if the OPDs for all points are vanishingly small, the lens may still not be usable in some coherent optical processor applications if there are appreciable OPDs between different object/image point pairs, resulting in large phase differences. Such lenses can still be used for processor applications where phase is not important.

For use in coherent processors where phase is important, standard CAD packages may be used to design lenses with better phase control and other characteristics if the correct procedures are used. Such lenses are generally known as Fourier lenses. If the lens is a Fourier relay lens, as described in Section 3.4, the procedure involves designing and optimizing the lens 'simultaneously' for good performance at two sets of conjugates: (i) infinite, corresponding to the plane wave Fourier components it must pass with low wavefront distortion (plane wave in, plane wave out) and near zero relative phase shift between the components, and (ii) finite, corresponding to the fact that it must perform good imaging, i.e., each input/output point pair must have a low OPD, and the image geometrical distortion must be sufficiently low. A low OPD of the chief (central) rays of all the plane waves is equivalent to a low relative phase shift between the plane wave components. This also must be made small enough to meet the processor requirement.

Automatic design

The general lens design procedure in packages with optimization, or automatic design capability, which includes all of the best packages and more recently some of the inexpensive ones, involves either manually setting up a 'first order' design from scratch or adapting an existing design, neither of which meet the performance requirement, but may meet it when optimized. The designer can then set up a 'merit function' for the lens based on the most important performance criteria. The simplest packages only permit geometrical ray-trace criteria such as spot size. The designer must also choose variables, i.e., those parameters in the lens that will be allowed to change during automatic design. If the designer is successful in picking a good starting point, choosing variables, and setting up the merit function, performance will usually improve rapidly with the first few cycles of optimization, and taper off. If satisfactory performance is not reached, a different starting design or choice of variables is necessary.

New features

New features are continually being added to these packages to make them more useful. For example, at least one package now has a 'diffraction modulation-transfer-function(MTF)-based' merit function and a 'user specified' merit function in addition to the standard geometrical ray aberration-based merit function. Standard optimization allows designs to approach a local mathematical optimum of the starting point design but not go beyond it. That is why the starting design must be changed when standard optimization does not give sufficient performance. This problem appears to have been eliminated by the introduction of 'global optimization'. This routine is known as Global Synthesis™ in the CODE V package. Global optimization allows designs to move out of the starting design region containing a local optimum and proceed to perhaps several new configurations that the designer may not have been aware of, which may have even better performance.

An important issue for optical processor design is the feasibility and desirability of designing and analyzing a complete system in a standard CAD package. In brief, it would be very desirable but is not yet feasible. It should be noted at this point that many optical processor systems may consist of several lens systems working together but performing different tasks with different requirements on each one. Sometimes a given lens will perform two tasks simultaneously for light in different directions or with different polarizations. Therefore, in general, lenses that perform a specific task need to be individually optimized for that task or tasks. After that is done it may be possible and desirable to enter the complete optical system to do certain overall performance analysis. This is not always a simple task. If the processor involves multiple optical paths (such as interferometric systems), these would have to be analyzed separately. If the system is a folded system with light passing in both directions through some of the lenses, it can sometimes be 'unfolded' for analysis in a standard package by entering some lenses, mirrors, etc. more than once. In general, entire system analysis, even when desirable, may not always be feasible.

Finally, all of the standard lens design and analysis packages have some limitations for the modeling and analysis of high-performance optical processors even if the systems can be entered in the package: (1) diffraction may not be adequately taken into account when using diffraction analysis options if the system is optically long, i.e. there is a large distance between 'pupils' containing many apertures such as SLMs; (2) the package may not permit the calculation of diffracted light levels with sufficient range to match the dynamic range requirement; (3) stray light levels cannot be modeled; (4) if compactness is important, '3-D' layouts of the complete system, including mounting hardware, spacers, baffles, etc. are very important, and for extremely critical systems, mounting hardware can have an effect on stray light; (5) many other additions to standard CAD packages could be suggested which would make them more useful for optical processor design, such as birefringent optical material capability to handle AO cells more accurately, polarization ray tracing for polarization-sensitive processors, easy calculation of response to a phase object such as a phase-modulated AO cell, and many others.

A partial listing of some of the standard packages available include ACCOS™ by Optikos, Cambridge, MA; BEAM3™ by Stellar Software, Berkeley, CA; CODE V™ by Optical Research Associates, Pasadena, CA; GENII™ by Genesee Optical Software, Rochester, NY; OPTEC™ by SCIOPT Enterprises, Sunnyvale, CA; OSDP™ by Gibson Optics, Sunnyvale, CA; OSLO™ by Sinclair Optics, Fairport, NY; SIGMA PC™ by Kidger Optics, Sussex, UK; and SYNOPSIS™ by BRO, Inc., Tucson, AZ.

3.3.2 Special CAD packages

Some special-purpose modeling and analysis packages which were developed for detailed modeling of stray light performance of large systems such as the Space Telescope, have been available for a number of years, running on mainframe computers: APART/PADE™ and GUERAP™ whose development began at the University of Arizona and Honeywell Aeronatical Systems respectively. The specialized training needed and high costs for use made them impractical for routine stray light analysis of small systems such as optical processors. APART/PADE takes a paraxial radiometric approach to evaluating stray light over user specified paths. The GUERAP™ program, which, at the time of writing, has evolved to GUERAP V™ and runs on 386/486 PCs, uses the statistical Monte Carlo method (with an efficiency enhancing technique) of tracing rays through an optical system. It is available from Lambda Research, Groton, MA. Another program called SOAR™, also available from Lambda Research, can calculate estimates of a number of types of scatter and diffraction in optical systems, including first order (two reflection) ghost images and narcissus, using point source transmittance. The present developer of APART/PADE, Breault Research Organization Inc (BRO), Tucson,

AZ has developed a general-purpose CAD modeling package, ASAP/ RABET™ which performs stray light analysis plus standard optical analysis. It has 3-D global coordinate entry for large, multiple path systems that can't be easily handled in standard optical design/analysis packages. Lens mountings, baffles, and other normally 'non-optical' surfaces can be included, as they can in the specialized stray light packages. Recently, some optimization capability has been added so that it is becoming more like a standard CAD design/analysis package. It is available for a number of systems, including 386/486 PCs. Other special CAD packages are available which allow mechanical layout and visualization, and some which also allow ray tracing. OPTICAD™ by Opticad Corp, Santa Fe, NM is one of the latter. A more complete list of scatter analysis packages can be found in Stover (1990, p. 161).

As discussed earlier, systems which make smaller performance demands on the optics will generally require less effort to design, especially if off-the-shelf lenses, etc. can be used. For high-performance coherent systems (high processing dynamic range, large TBW product, low phase noise, etc.), especially when coupled with a need for good thermal stability or vibration immunity, more attention to the details of the optics and the mechanical design will be required. When off-the-shelf lenses do not meet the requirement, which will usually be the case for high-performance systems, a lens designer familiar with Fourier or athermal design may be needed to assist the system designer.

If lens design or overall systems design needs to be made athermal, at least one standard package, CODE V, permits athermal lens design, i.e., temperature is a system variable. However, the authors are not aware of any system *analysis* package at this time, including ASAP, which permits optical analysis over temperature. Hence, if the complete system cannot be entered into CODE V it will require some 'manual' thermal calculations followed by reentry into the optical analysis package to determine when the system is sufficiently athermal. Similarly, analysis of the effects of vibration on the performance of a lens can be done in at least one optical package (CODE V) using the automatic tolerancing routine. Vibration amplitudes can be entered into CODE V via a NASTRAN interface from a mechanical model of the optical system (private communication, Kevin Thompson, ORA). If the resultant phase dither, resolution degradation, etc. is too large based on the CODE V calculation, either the optical sensitivity or the mechanical design must be modified, or else the vibration input must be attenuated further. The usual sequence of attack, in order of increasing difficulty, is (1) external attenuation, (2) mechanical rede-sign, and (3) optical redesign. In some cases one or more of these may not be an option.

3.3.3 Mathematical simulation packages for signal processing

Although a desirable goal, it is still not possible to simulate the performance of a particular optical processor design completely using a single optical CAD package in the sense that one can input signal data and obtain a calculated version of what the real processor output should be. At present it is necessary to calculate basic performance data of an optical design using the optical CAD packages already described, and then to transfer the basic performance data into a suitable mathematical processor model which has been set up in one of the mathematical simulation packages. In other words, the desired basic optical performance of the processor design, such as spot size or resolution, channel crosstalk, phase and amplitude distortion, etc., can be entered along with the desired signal data, into the mathematical processor model and the effect of one or more basic performance factors can be calculated. In the models set up so far by the authors, it is natural to model both phase and amplitude errors simultaneously. Geometrical distortion, when the effect is amplitude error, can also be included. More detailed models which include variation in resolution and optical crosstalk should also be possible.

Mathematica^{TM}, DADISP^{TM}, and Prime Factor FFT^{TM} are exam-ples of packages that can be used for signal processing purposes. Prime Factor FFT, available from Alligator Technologies, Fountain Valley, CA, is a library of callable linear data processing functions such as 1-D and 2-D transforms and filters. It supports various popular program-ming languages. DADISP, on the other hand, is a self-contained general-purpose data processing program that can be used to generate waveforms and perform many different transforms, filter, and matrix operations. It can display a number of signal waveforms corresponding to different locations in a processor in a windows format. It is available from DSP Development Corp., Cambridge, MA. Mathematica is also a self-contained program, but functions like a high-level programming language with callable routines. It contains many general-purpose mathematical routines, some of which are useful for signal processing, as well as symbolic math and integration. It is available from Wolfram Research Inc., Champaign, IL.

No matter which package is used, the general procedure is to simulate the operation of a particular processor, first with 'ideal' optics and signals (no errors), and then with 'real' optics and signals, with the errors introduced as described above. The degraded performance can then be compared with the ideal case, giving an estimate of the degradation. This may consist of signal loss (sensitivity change), peak/sidelobe ratio changes for correlation, distortion, or other effects depending on the processor and model. Examples of using Mathematica to simulate the effects of phase errors will be given in Section 3.5.

3.4 Constraints due to physical optical phenomena, design, and fabrication

3.4.1 Physical optical constraints

Diffraction, crosstalk, and dynamic range

Even if a particular optical design could be implemented perfectly, there would still exist limitations due to fundamental physical phenomena. These include, primarily, diffraction and stray light. The laws of diffraction determine the limiting or ideal spatial energy distribution in an optical channel, based on the details of the optical design and the light source being used. Stray light has historically been treated as an addition to the ideal diffraction solution when it is treated at all. In an optical processor, where one processes many channels in parallel, the dynamic range of the processor is reduced over that of a single-channel processor by crosstalk between neighboring channels by the spreading and overlap of the diffraction 'tails' as discussed in Section 3.2.

As a specific example of optical crosstalk effects in a processor, consider the operation of the optical rf spectrum analyzer, or rf channelizer. (Performance examples are discussed in Section 3.5.) The amplitude of the optical Fourier transform of the signal in the input aperture is detected by an array of photodetectors at the output/transform plane, and the parallel output of the array is equivalent to a bank of contiguous, narrow-bandpass filters. An actual implementation could, in fact, consist of electrical filters such as lumped-element circuits or SAW filters, F1 through FN, as shown in Figure 3.1(a). The parallel, electrical outputs of the filters are directed to threshold detectors or other post-processing circuitry. A schematic of the optical

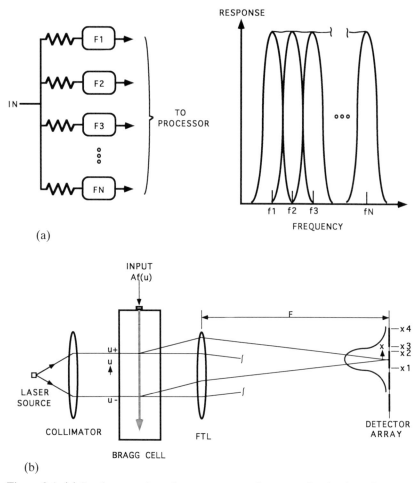

Figure 3.1 (a) Implementation of a spectrum analyzer as a bank of contiguous, narrow-bandpass filters. (b) Schematic of 1-D AO power spectrum analyzer.

implementation is shown in Figure 3.1(b). The spot intensity profile at the detector array due to a signal $Af(u)$ in the input aperture is just the square of the Fourier transform of the signal in the input aperture, which extends from $u-$ to $u+$. The spot is shown centered on one pixel, and for this case the adjacent channel crosstalk is the integral of intensity from x3 to x4 divided by the integral from x1 to x2. It is beyond the scope of this chapter to discuss all the various issues in the implementation of a good channelizer rf receiver; these may be found in many references (Anderson, Webb, Spezio & Lee, 1991). However, we shall concentrate on the one critical specification of such a receiver – the channel bandshape, or single-frequency response for each of the channels. The response in the case of lumped-element filters is dependent on the number of poles in the filter design; in the case of SAW filters, the design of the electrical-to-acoustic transducer is the determining factor.

However, for the AO processor the channel response depends on very different physical phenomena. Here the single-frequency response shape is determined by (1) the input optical beam amplitude profile (apodization) and wavefront distortion (beam quality), (2) unwanted acoustic–energy density variations along the illuminated cell aperture, (3) the lens/detector design which determines the basic diffraction plus ghost image response shape, and (4) small-angle scattering of the diffracted beam, due to imperfections in fabrication and materials. In order to estimate the single-frequency intensity profile in the frequency plane, a Fourier transform calculation can be performed, using as inputs careful measurements of the amplitude profile of the optical beam and the phase errors introduced by the light source and the optical elements as the light passes through. However, such a calculation is generally inadequate, as discussed in Section 3.5. Aside from the fact that continuous, subtle, diffraction effects are ignored throughout the optical system when a ray trace/Fourier transform is used, stray light and optical scatter in a multi-element optical system are not explicitly handled when this calculation is performed by a standard package such as CODE V. That is, the effect of ghost images is not included in the diffraction calculation (it is a separate geometrical routine), and there is no way to include the effect of scattered light.

Scattered and reflected light

Crosstalk in an optical system is fundamentally determined by optical diffraction limits, and secondarily by stray light-scatter and unwanted reflections, i.e., ghost images. All are, to a degree, under the control of the designer, who can influence whether diffraction, scatter, or ghosts, or what combination of the three will ultimately limit performance. However, if basic design and fabrication are sufficiently good that diffraction is not the limiting factor, scatter and/or ghosts will normally be the limiting factor.

As an example of stray light analysis to predict crosstalk for a real optical system, Figure 3.2 shows the calculated light distribution at the

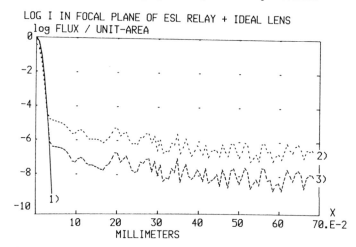

LOG I IN FOCAL PLANE OF ESL RELAY + IDEAL LENS

Figure 3.2 Calculated light distribution vs angle at the image or Fourier plane of a 24 surface Fourier relay lens. The span, 0.7 mm, corresponds to 0.1°: (1) no scatter, no ghosts, (2) 50 Å rms roughness, 4% reflectivity, (3) 5.0 Å rms roughness, 0.4% reflectivity, all surfaces.

image or Fourier plane of a Fourier relay lens (similar to the one in Figure 3.6) followed by a Fourier transform lens. The package used for the analysis was ASAP. The light illuminating the system is made nearly ideal for this analysis so that the effects of stray light in the optical system can be clearly seen. This is done by letting the light be Gaussian-shaped, single-frequency, and 100% coherent, similar to what can be obtained from a single-mode fiber-coupled laser with a suitable mask to remove the sidelobes. Thus diffraction sidelobes are not generated and crosstalk performance is determined only by stray light. Also, the transform lens is an ideal lens (available in ASAP and some other optical CAD packages) and does not contribute any degradation to the system. Thus degradation is due entirely to the relay lens. With perfectly smooth surfaces and zero reflectivity, the calculated focal spot intensity, due only to diffraction, is curve (1). Curve (2) is for the case when all surfaces have 50 Å rms roughness and 4% reflectivity, equivalent to a 'normal' polish for glass and no anti-reflection (AR) coatings. Curve (3) is for 5 Å rms roughness and 0.4% reflectivity, equivalent to an unrealistic 'superpolish' for glass and average AR coatings. Curve (3) also assumes no scatter from coatings. In reality coating scatter can be expected to be higher than this.

Actual measured performance (sidelobe levels) would be expected to be somewhat higher than curve (2) because (*a*) the real laser will not be perfect, (*b*) the fabricated optics will not be as good as the design, (*c*) scatter theory is still not perfect, and (*d*) absorption, dirt, and other anomalous factors will be present. Nevertheless, these calculations are very useful for predicting the 'maximum possible performance' bound.

Finally, note that curves (1) and (2) and (1) and (3) cross. The decrease in area under the curve inside the crossover (actually a volume in space) should equal the increase outside, and represents energy scattered and reflected into the sidelobes. In general, in an optical system with more than two optical surfaces, which is almost always the case, multiple ghost images can create irregular plateaus on the intensity distribution plot while the scatter will be a smoothly changing function. For a large number of ghost images, the phases will tend to be random at the image plane of interest, and there will tend to be no 'dc' components, i.e., the effect of ghost images will be to modulate spatially the scatter distribution. In Figure 3.2 there are all possible combinations of 24 surfaces contributing ghost images, i.e., 276 images, which combine to give the irregular distribution shown. With ghost images 'turned off', scattered light would have the greatest effect on nearest neighbor crosstalk and dynamic range for detector pixel sizes approximating the 'spot size', because scatter alone normally rises smoothly with decreasing angle. For the irregular distribution of scatter and ghosts combined, this will not always be the case. Curve (2), which is for a 'normal' polish and no AR coatings, indicates that the stray light near the diffraction spot (about 0.007°) cannot be any less than -50 dB below the peak. Integration over pixel geometry and other factors will further lower the dynamic range prediction to 40 dB or less. Curve (3), which is for a less than optimum AR coating reflectivity but an unachievable 'superpolish' for glass and coatings, predicts a 15 dB improvement over curve (2). While somewhat unrealistic, this example shows how the effects of stray light can be examined in a processor design, and how stray light can limit dynamic range when the resolution is near 0.01°.

Therefore, when it is necessary for an optical processing system to have diffraction-limited resolution and large dynamic range (maximum sidelobe suppression), it is important to consider stray light, particularly small-angle scatter. If the scattering function of each element or surface in a system is known or can be estimated, and a special analysis package such as ASAP is used to supplement the standard CAD design package, it is possible to include the effect of stray light performance during system design. This will allow stray light effects to influence the design, just as ray trace and diffraction calculations normally do. The designer will be able to see the effect of such factors as surface polish, glass homogeneity, and coating scatter which are included in measured scatter, and the effect of the number and placement of elements on scatter, or other design changes. This has already been done on a limited scale for some simple optical arrangements and some common AO processor crystals and glasses (Brown, Craig, & Lee, 1989). Therefore, we present the following brief description of optical scatter.

Scatter measurement

The optical scattering function for transmissive optics may be expressed in a manner independent of any particular optical measurement system by using the bi-directional transmission distribution function (BTDF), which has been defined for measurement purposes as (Dereniak, Brod and Hubbs, 1982)

$$\text{BTDF} = V_s(\Theta_s)/V_{ns}(0)\,d\omega\cos\Theta_s \qquad (3.1)$$

where Θ_s is the scattering angle, $V_s(\Theta_s)$ the detector signal at Θ_s, $V_{ns}(0)$ the detector signal at $\Theta_s = 0$ with no sample, $d\omega$ the solid angle of detector and $\cos(\Theta_s)$ the obliquity factor. BTDF is directly related to a similar function, BRDF, which describes purely reflective surface effects (Nicodemus, 1965). However a BTDF measurement of a lens or Bragg cell crystal will include scatter from both surfaces, from imperfections and inhomogeneities in the bulk material, and from any residual subsurface damage due to grinding and polishing. Hence a BTDF curve for a standard thickness and surface quality of an optical material, and especially if a scalable definition were possible, would be a valuable piece of design information. Nevertheless, whenever BTDF data exist for a material in an optical system, half the value can be assigned realistically to each surface of a lens element or Bragg cell when using the RABET[TM] portion of the ASAP package. This is because scatter is assumed to occur only at surfaces in RABET. If BTDF data exist for all the surfaces in the system, including baffles and other 'nonoptical' surfaces which may be painted a flat black or designed deliberately to scatter any light hitting them, RABET can give the combined effect of all scatter on system performance.

Note that the angular resolution of the BTDF depends on $d\omega$, the solid angle of the detector as seen by the sample. If dω is made small, say 0.001° in the measuring plane, measurements can be made with this resolution for a small range near $\Theta_s = 0$, where the light intensity is still measurable. In order to continue measurement to large angles dω must be gradually or periodically increased, which can introduce errors in the results. In principle, for infinitesimally small dω, the true BTDF may be measured at any angle down to $\Theta_s = 0$, and will be independent of the sample size and detector size. However, in practice there are many obstacles to overcome, a major one of which is that there is a 'crossover' of the scattering function of the scatterometer with no sample inserted (the signature) and the measured data with a sample present. Since the signature must be subtracted or deconvolved from the measured data to give the BTDF of the sample, there is a region of uncertainty near the crossover where the values are equal within the measurement uncertainty, and BTDF cannot be obtained by simple subtraction. It can be seen that reducing the instrument signature permits BTDF measurement to smaller angles, since the crossover will occur at a smaller angle. For the same reason, more highly scattering samples can be measured to smaller angles. Another important factor increasing the difficulty of small-angle BTDF measurement is that after the signature is measured, and the sample is inserted, the diffracted light distribution is changed by the sample as well as the scattered light distribution. Also the sample reflects and absorbs light as well as scattering it, and corrections must be made for all of these. Details of this and many other aspects of scatter measurement can be found in Stover (1990).

Finally, it is important to note that the BTDF depends on the optical wavelength. Scatter levels in glass due to artifacts smaller than the wavelength (pure Rayleigh scatter) should vary with wavelength as $1/\lambda^4$, but Rayleigh scatter is typically very small for good optical materials. Scatter by artifacts equal to or greater in size than the wavelength usually predominate, and the variation with wavelength will be less than for pure Rayleigh scatter. Major sources of optical scatter include surface finish, subsurface damage, bulk particulates, index fluctuations, and surface contamination. Although increasing the wavelength will, in general, reduce scatter levels, it also will reduce the diffraction efficiency and diffraction angle of AO cells, as well as increase their design difficulty. Hence a trade-off between these factors can more easily be made as new laser types become available and available power increases.

BTDF measurement

Very-small-angle BTDF data were measured at TMA, Inc., Bozeman, MT., for a number of samples of AO materials – gallium phosphide (GaP), lithium niobate(LN), tellurium dioxide(TeO$_2$), and a material under development, mercurous chloride(HGCL), as well as some glasses, and reported in Stover, Bjork, Brown, & Lee (1990). These data were measured with a newly designed scatterometer and 860 nm laser source, at two linear polarizations, using highly polished and characterized samples, with the measurement range extending from 0.0001° to 80° in each direction. However, reliable data could not be obtained down to the 0.001–0.002 degree-from-specular value desired for high-resolution TeO$_2$ processor applications.

Figure 3.3 shows the results of one measurement on a sample of TeO$_2$, 8 mm thick, intended for use as a Bragg cell, and a glass for comparison. Results of small-angle scatter for this and other samples, including glass, ranged from about 10 for glass, 20 for TeO$_2$, 30 for GaP, 100 for fused silica, and 1000 for LN and HGCL at 0.1° to 10 000–900 000 at 0.01° for crystals only. Scatter of glass could not be determined at 0.01° due to the crossover occurring at that angle. Although the crossover of some measurements was less than 0.002°, values for less than 0.01° may not be reliable. The possibility of reliable measurements extending down to 0.001° will depend on improvements in low scatter mirrors, scatterometer design, detection methods, and deconvolution of the signature from the data.

Direct performance limits and small-angle BTDF

If the BTDF of the TeO$_2$ shown in Figure 3.3 were the only source of scatter in a spectrum analyzer with 0.1° channel spacing, it would have over 60 dB dynamic range due to scatter crosstalk. With 0.01° channel spacing, it would have little more than 20 dB dynamic range due to scatter crosstalk. As previously mentioned, it would be desirable to know what the BTDF and scatter crosstalk would be for 0.001° channel spacing. Short of being able to measure BTDF to the desired small angles, or actually constructing the spectrum analyzer itself, it may be possible to perform modeling that extrapolates the observed scatter function down to much smaller angles (Brown, Craig, & Lee, 1989).

Small-angle BTDF prediction

Much work has already been done by many people to analyze one major contributor to the BTDF – surface roughness. Unfortunately,

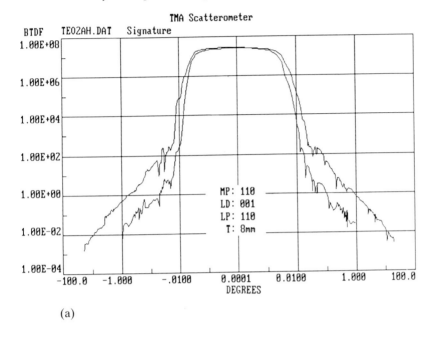

(a)

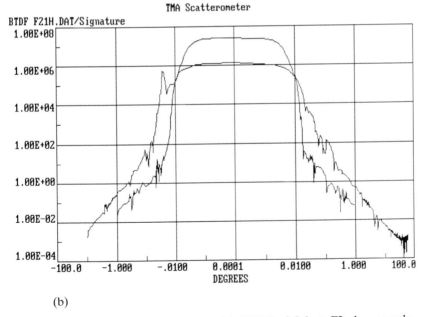

(b)

Figure 3.3 (a) BTDF of TeO$_2$ sample. (b) BTDF of Schott F2 glass sample. Both are for 860 nm.

good agreement between the models and measurement has not yet been obtained. Nevertheless, we present here a few basics of a 1-D scalar model. If a surface profile is measured accurately over an aperture D, and if the aperture is fully illuminated, one should be able to calculate scatter down to an angle $2\lambda/D$, since it requires a minimum of two cycles of the longest *spatial* wavelength present ($\Lambda = D/2$) to diffract light of wavelength λ. For high-quality surfaces the amplitudes of the Fourier components of surface roughness, a_n, are much less than λ, and for this case the intensity of a particular component, I_{nm}, is given by Eq. (3.2) (Bennett, 1978; Church & Zavada, 1975).

$$I_{nm} = I_0(2a_n\pi/\lambda)^{2m}\cos\Theta_i\cos\Theta_{snm}, \tag{3.2}$$

where I_0 is the incident intensity at angle Θ_i, λ is the optical wavelength, and Θ_{snm} is the scattering direction of the nth Fourier component, mth order, which is determined by the grating equation,

$$\sin\theta_{snm} = \sin\theta_i \pm mf_n\lambda. \tag{3.3}$$

Here f_n is the nth spatial frequency component and m is the diffraction order. For micro-rough surfaces $a_n/\lambda \ll 1$, and values of $m > 1$ can be ignored.

In order to attempt to apply 1-D scalar theory, a means of measuring the surface profile over a sufficiently large aperture is required, and we have traded one difficult measuring task for another. Fortunately, noncontacting, optical phase-shift interferometers are available which can measure the amplitude of micro-roughness and the spatial-frequency spectrum of the rough features over a limited range using a 1- or 2-D focal plane array. However, when the magnification is set to give good spatial resolution over the sample, the range is too small. By combining a number of measurements using special software for that purpose, a total scan length of 17.3 mm was obtained for TeO$_2$. Spatial frequencies with wavelengths down to 17.3/2 mm should produce scatter at angles down to 0.0005°. The 'longscan' measurement, performed by E. R. Cochran at Wyko Corp., Tucson, AZ., is shown in Figure 3.4. The rms roughness of the surface was 1.03 nm for this sample. The measurement was made for the scattering direction of greatest interest, i.e., the optical diffraction direction for a shear-mode on-axis cut TeO$_2$ Bragg cell, and was used by Breault Research Organization, Tucson, AZ., to extrapolate measured BTDFs to smaller angles (Brown, Craig, & Lee, 1989). The accuracy at very small angles, of course, cannot be checked by measurement, but if it could the measurement would be combined with scatter from the other

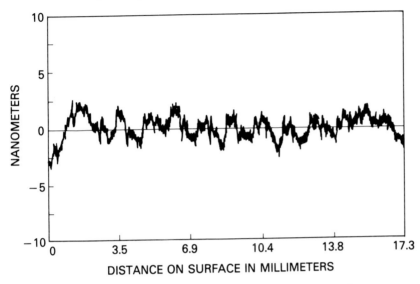

Figure 3.4 1-D surface roughness profile of a TeO$_2$ sample.

contributors to the BTDF (particulate, inhomogeneity, subsurface damage, etc.) as well.

Because surface roughness and scatter are 2-D phenomena, 2-D vector models may be necessary to relate them adequately. With an adequate 2-D model and a method to splice together 2-D surface data, it should be possible to extrapolate surface-produced BTDF to small angles more accurately. Also, this model will need to be combined with models for the other known contributors to BTDF.

Ripple

The residual error remaining in a system in spite of good design and fabrication will, of course, constrain the performance through diffraction. Estimates of the combined effect of both amplitude and phase errors in an optical system can be obtained at any time during or after the design process using any standard CAD package with fast Fourier transform capability, as discussed throughout this chapter.

However, sometimes it is desirable to be able to make preliminary estimates of the effect of errors of a particular type in optical processors without performing any design or calculation. When the error can be represented or approximated by a standard mathematical function, and the process involves standard transform operations, the estimate can often be obtained using standard tables of transform pairs.

A particularly serious type of error is frequency-response ripple and phase ripple. Their effect has long been a concern in pulse-compression radar systems. The effect of electrical amplitude and phase errors in pulse-compression radar matched filter design is analogous to the effect of optical amplitude and phase errors in optical correlators. Constant amplitude ripple in the frequency response of the radar matched filter was observed to create a pair of 'echoes', or sidelobes in the time response. Constant phase ripple was observed to give a similar result, with a time response given by

$$a_n[g(t + n/B)]/2 + g(t) - a_n[g(t - n/B)]/2, \qquad (3.4)$$

where a_n is the magnitude of the phase ripple in radians, $g(t)$ is the undistorted signal, n is the number of ripples, B is the bandwidth, n/B is the separation in time between the signal and the echoes, and the contributing phase distortion given by

$$G(f) \exp [ja_n \sin (2\pi nf/B)] \approx G(f)[1 + ja_n \sin (2\pi nf/B)] \qquad (3.5)$$

where $G(f)$ is the Fourier transform of the undistorted signal and the approximation can be used when ripple is small (Farnett, Howard, & Stevens, 1970). These formulas can be used to estimate quickly the effect of phase ripple in an optical correlator. If the effect is greater than the peak sidelobe level of the weighted signal (which can be -43 dB for Hamming weighting or even lower for $\cos^4$ or Gaussian weighting (Harris, 1978)), ripple will limit system performance. The above relationship indicates that if the magnitude of the phase ripple is $1°$, the echo level will be -40 dB. In fact, a rule of thumb, stated in Farnett *et al.* above, is that electronic filter designers usually associate a -40 dB echo level with amplitude and phase tolerances of 0.1 dB and $1°$. Such tolerances are extremely difficult or impossible to achieve in optical systems. Fortunately there is usually no tendency for ripple components to occur when fabricating conventional optics such as spherical or plano surfaces, but they can occur in some computer-controlled polishing operations on aspheres. Phase ripple can also occur due to index ripple in some optical glasses, resulting in phase ripple on the order of $\lambda/100$, or $3.6°$ (private communication, Optical Research Associates, Framingham, MA.). Even higher levels of index striations have been reported in some types of fused silica. Ripple is, of course, related to scatter since a particular frequency of ripple is equivalent to a larger than normal Fourier component of surface roughness or other scatter contributing factor. Hence, phase ripple due to these or other causes must be carefully controlled to avoid limiting dynamic range.

For an AO Fourier transform device a 5° criterion corresponds to an absolute wavefront accuracy of $\lambda/72$ over a uniformly illuminated aperture. Such absolute accuracies are unusual but not impossible; microscope objectives routinely achieve $\lambda/50$ accuracies for broadband incoherent light, and collimators for semiconductor laser diodes approach this performance for coherent light (Forkner & Kuntz, 1983). However, in interferometric processors (Section 3.4.2) it is the relative optical phase accuracy between the signal and reference light beams that is relevant. Common-path and almost-common-path interfero-meters have demonstrated the requisite degree of phase accuracy (Berg, Lee, Casseday, & Katzen, 1978).

3.4.2 Design constraints

Good performance in 1-D or 2-D optical processors, which may involve multiple simultaneous and/or sequential Fourier or matrix operations on one or more arrays of fixed or moving complex data (i.e., information residing in both amplitude and phase) which are modulated onto light beams, obviously depends heavily on the optical architecture and design, and how well they control optical errors and influence physical optics constraints. Since there may be more than one design capable of performing the same process, each with different advantages and limitations, design itself can be considered a constraint. Some of the most basic design factors which limit what can be achieved will now be discussed.

Fourier optics

Traditionally, coherent optical processors have employed a '4F Fourier system' consisting of at least two high-quality objective lenses to transfer amplitude and phase information between three planes, labeled input plane, Fourier transform or 'filter' plane, and output plane in Figure 3.5. With such an arrangement, the input-to-output distance is fixed at approximately 4 × focal length. Various 1- or 2-D SLMs, AO cells, photodetector arrays, or nothing at all, may be located at the various planes. With the input plane at the front focal plane, the transform plane at the back focal plane of the first lens, and the inverse transform plane at the back focal plane of the second lens,

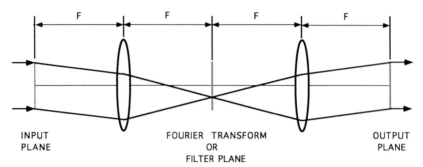

Figure 3.5 Basic '4F' imaging system consisting of two Fourier transform lenses of focal length F.

this basic processor is free from phase curvature (quadratic phase error), for paraxial light (light traveling very nearly on-axis), as shown by Goodman (1968, p. 85). Many 'architectures' consisting of variations and combinations of the basic processing unit have been investigated for various applications. In some, the output plane of one basic unit may be the input plane of a following unit. In other 'time integrating' architectures (Casasent, 1983) data may also be modulated onto all optical channels in parallel by modulating the laser source. Or, the processor may work entirely in the time domain with no active device at the frequency or transform plane, i.e., the AO 'space integrating' correlator example, discussed later.

In many present-day processor architectures the effective focal length F is often required to be large in order to match existing SLMs or detector arrays. The traditional approach, even if there were no other problems, may result in a basic processing unit exceeding a meter in length, and more than one unit in tandem may be required to perform the necessary processing. A long system tends to be mechanically unstable and affected by temperature and vibration, even when folded to shorten its length. In addition to compactness, another problem encountered in using simple objective-type lenses is that they cannot give good imaging performance and low phase error over a large field. A compact processor with a high bandwidth requirement, i.e., large angular field optics, and/or a low-phase-error requirement may exceed the capability of the traditional approach. If a system is interferometric, i.e., has multiple beam paths, and uses phase modulation, it may have a low-phase-error requirement. For these and other reasons, optical processing systems which are intended to be compact, mechanically stable, transportable, or suitable for installation in environments such as aircraft, usually cannot be built using simple doublet lens design.

Fortunately it is possible to design more compact, optically corrected Fourier lenses which have good control over the variation of optical path length, or phase error, using optical CAD methods. Using the telephoto principle allows a significant reduction in the input/output distance while maintaining a long effective focal length. An early design effort in this area resulted in shrinking a $4F$ system to $1.4F$ while maintaining $\lambda/8$ (eighth wave) phase error (Blandford, 1970). This design made use of the telephoto principle. The number of aberrations to be controlled (besides phase error) determines the necessary degrees of freedom, which, in turn, sets the minimum number of lens elements required. It is possible to design Fourier lenses with nearly zero front and back focal distances, reducing wasted space to a minimum. The lenses themselves do occupy significant space and may need 8–12 elements for a Fourier pair or a basic processing unit. They can be very expensive to design and fabricate 'from scratch' if the required phase error is low – less than $\lambda/8$ – and is coupled with large angular field requirement.

Fourier relay lenses

When there is no operation to be performed at the intermediate plane, a pair of telephoto Fourier lenses serves only to relay a wavefront accurately from the input plane to the output plane, and will be referred to as a Fourier relay lens (FRL). The use of one or more high-performance FRLs in a system requires serious consideration due to the cost, size, and weight, each of which can be a constraint. For a 30 mm input/output format, an angular field of 9°, effective design peak-to-valley wavefront distortion of $\lambda/15$ with possibly $\lambda/10$ in the finished lens, the cost could easily exceed $100 000 (private communication, Optical Research Assoc., 1993). Off-the-shelf Fourier transform lenses (half of a FRL) are available at this time from at least two sources, one Japanese and one US, but the choice of models is very limited. A layout drawing of a current design for an FRL intended for use in an AO space-integrating correlator, is shown in Figure 3.6. It was designed by Jan Hoogland, Grants Pass, OR., to NRL specifications. The first and last elements are actually the portions of the Bragg cells between the acoustic beams.

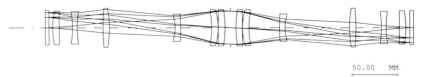

50.00   MM

Figure 3.6 1:1 Fourier relay lens layout, designed for ± 5° operation, 830 nm wavelength, $\lambda/16$ or less peak-to-valley wavefront error in design, $\lambda/7$ or less in the finished lens, and optimized for use with 8 mm thick TeO_2 shear Bragg cells. Designer: J. Hoogland.

Another viewpoint, that of standard plane wave Fourier analysis, is useful for analyzing the operation and phase relationships in the FRL (and consequently the imaging correlator discussed later). A modulated or diffracted wavefront at the front focal plane can be decomposed into, or considered to be, a unique set of plane wave Fourier components. Each component has a unique direction, amplitude, and absolute phase, which must be preserved while being transmitted by the relay lens. When the components are reassembled at the back focal plane (in the second Bragg cell) into the transmitted wavefront, it will resemble the initial wavefront if optical errors are kept low. For a perfect FRL:

(1) All plane wave components remain plane – OPD is zero for all plane waves.
(2) A plane wave component entering a lens has the same (but reversed or complementary) direction when exiting.
(3) All plane wave components have the same time delay or absolute phase, i.e., no frequency-dependent phase error; this is equivalent to (4).
(4) The chief ray OPD is zero for all plane wave components.
(5) All plane wave components have perfect registration at the output plane – this is equivalent to (6).
(6) A point centered in the input plane is perfectly imaged to a point centered in the output plane – this is a consequence of (4).
(7) All other points in the input plane image to points in the output plane at the correct *x, y* position – this is equivalent to (8).
(8) Magnification is unity and geometrical distortion is zero.
(9) All plane wave components are passed by the lens – there is no cutoff.

All this means is that the modulated wavefront at the input plane is reproduced at the output plane with no error.

Based on the above, a brief sketch of how a Fourier lens design would be carried out by a designer was given in Section 3.3. However, the actual procedure would vary considerably depending on the specifications. FRLs are often designed assuming the input and output planes are in air, even though they may actually be in a refracting medium, as they are in most AO processors. In the correlator, part of each AO cell is in the interferometer path and, ideally, should be treated as an element of the relay lens. The primary effects of not doing this are spherical aberration (SA), i.e., wavefront distortion (WFD), and a relocation of the focal planes. This happens even if the Bragg cell crystals are optically perfect. Like the nonimaging correlator phase error described by Eq (3.7), SA depends on the thickness and angular range, but unlike nonimaging phase error, SA can be partially reduced by refocus (Smith, 1966, p. 294). In processors where the angular range (bandwidth) is low so that SA is small, it may be possible to use an FRL designed and optimized for use in air. Even so, it will be easier to achieve the required system performance when the design is as accurate (realistic) as possible.

Binary optics

Because of the above considerations an alternative to conventional glass lenses is very desirable for optical signal processing. One potential alternative is binary optics, a form of holographic lens. Ordinary holographic lenses are made by exposing a photosensitive material to a suitable optical interference pattern, and processing the result as for a hologram. Ordinary holographic lenses are limited in phase accuracy and efficiency by materials-related problems such as dimensional changes, and the difficulty of generating a sufficiently accurate interference pattern. A more precise and promising type for optical processing is binary, which is computer-generated and 'written' onto relatively thin and therefore lightweight glass substrates. The process uses some of the techniques developed for electronic integrated circuits. One or more processing masks are used to generate steps in the optical surface, causing it to approach more closely the shape of an ideal optical surface, which is not restricted. A lens created with one mask has only two levels or phase values, while a two-mask lens has four phase values. Diffraction efficiency increases with the number of levels and is close to 100% for 16 levels (four masks), but alignment difficulty also increases with the number of masks. The difference in optical path for adjacent steps in a two-level binary lens is one wavelength, for a four-level one it is a quarter wave, etc. Thus a simple binary lens is inherently a single-wavelength device, but can function in broadband applications when combined with refractive elements (Faklis & Morris, 1991). Binary optics are under development at several locations in the US (see Veldkamp & McHugh (1992), S Lee (1990), and Buralli & Morris (1989)). A conceptual drawing of a binary-type Fourier relay lens is shown in Figure 3.7. It uses two diffractive lenses of a type described by Buralli & Morris (1989). The number of levels or steps used in fabricating each binary element is not shown. A successful binary FRL design may require more than two diffracting elements, possibly four. Compare this with the twelve standard elements required for the conventional Fourier relay lens in Figure 3.6.

1-D AO space-integrating correlator operation

This processor will serve to illustrate a number of design constraints common to 1-D AO correlators. First, we describe its operation. The basic optical layout is shown in top and side plan views in Figure 3.8, where two AO modulators or cells are shown illuminated by a collimated beam of light. Each AO cell contains an acoustic traveling wave generated by a piezoelectric transducer driven by an rf signal modulated with the phase and/or amplitude information to be correlated. The modulated traveling wave in the first cell diffracts light proportionally to the rf power, thus transferring the instantaneous phase and/or

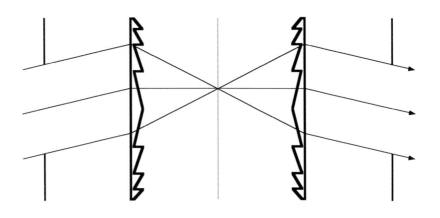

INPUT PLANE   DIFFRACTIVE FOURIER RELAY LENS   OUTPUT PLANE

Figure 3.7 Conceptual binary diffractive Fourier relay lens formed by combining two telecentric paraxial diffractive lenses, described by Buralli & Morris (1989).

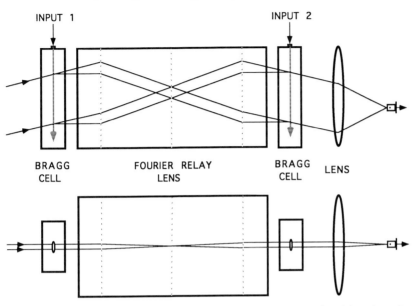

Figure 3.8 1-D AO space integrating correlator using conventional imaging. A single diffracted component and the undiffracted component are shown.

amplitude information onto the diffracted optical beam. The modulated, diffracted beam is imaged by means of the FRL into the second cell where it is diffracted a second time by the other signal to be correlated.

If both signals are identical, at the instant of peak correlation all plane wave Fourier components of the doubly-diffracted signals are perfectly aligned with the reference wavefront due to the symmetry of this architecture. Symmetry is made possible by operating both cells in the same diffracting mode, either upshift or downshift. The first AO cell performs the function of a grating beam splitter for all the diffracted Fourier components of the signal and for the undiffracted reference beam, while the second AO cell performs the function of a grating beam combiner. Hence this type of correlator can be considered an interferometer in which the input AO cell functions as a grating beam splitter and the output AO cell functions as a grating beam combiner. If the FRL did not have to pass the undiffracted reference beam, the angular working range of the FRL could be smaller (reducing the cost and weight of the FRL), but the penalty would be increased mechanical/phase instability because the interferometer would no longer be nearly-common-path. This conceptual

'Mach–Zehnder' arrangement is shown in Figure 3.9 with the undiffracted reference beam directed by two prisms around the lens, recom-
bining at the second Bragg cell. The lens following the second cell,
which is outside the interferometer section and is not critical, collects
and focuses all light across the optical aperture (space integrating). The
single detector is placed (ideally) where plane wavefronts focus, giving
maximum light in the detector at the correlation peak and minimizing
detector size. The detected heterodyne signal contains Fourier components of the doubly-diffracted signal shifted in frequency by the sum of
the two rf carriers.

Finally, since the relayed information from the first cell propagates
in the opposite direction from the signal in the second cell, the natural
output of the system is the convolution of the two signals; if one of the
signals is time-reversed as shown in the figure, the photodetector
output is, ideally, the cross correlation (CrossCor) of the two signals,
given by either Eq. (3.6) or (3.7). Here FFT is the fast Fourier
transform of the complex-valued list

$$CrossCor(list1, list2) = IFFT[FFT(list1)\ Conj\{FFT(list2)\}] \qquad (3.6)$$

$$CrossCor(list1, list2) = \sum list1(n)\ list2(n + m) \qquad (3.7)$$

of signal samples, IFFT is the inverse FFT, and Conj is the complex
conjugate. Eq. (3.6) represents the standard algorithm for obtaining
the cross correlation of two lists of complex data of equal length – see,
for example *Fourier Perspective II, Users Manual*, Alligator Transforms, Costa Mesa, CA. Correlation, however, does not normally
require the frequency domain to describe its operation, i.e., both the
inputs and the output are in the time (or space) domain, and can be
carried out strictly in that domain, as is done in the equivalent
algorithm described by Eq. (3.7). Here m is the shift variable describ-

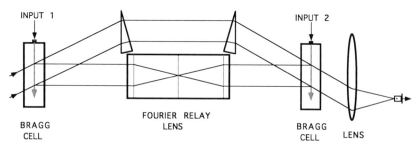

Figure 3.9 Conceptual 'Mach–Zehnder' 1-D AO space integrating correlation.

ing the translation of list1 with respect to list2, and the sum is over n, the length of the list. The transform method, described by Eq. (3.6), is normally used to compute cross correlation because of speed. It also permits access to the frequency domain. However, the model described by Eq. (3.7) can actually be faster when modeling space integrating AO correlation with fixed, position-dependent optical error, and it closely resembles the actual correlator operation, but it does not permit access to the frequency domain. Further discussion and results of error effect modeling will be found in Section 3.5.

Proximity imaging correlator

In a slightly different version of this correlator, as described by Torrieri (1983), the two Bragg cell acoustic beams are close together and conventional imaging is not used. This results in unwanted phase errors due to light propagation between the first and second acoustic beams. This technique is sometimes called shadow-casting or proximity imaging. In the diffracting plane, the phase error function, $E(\theta)$, is a quadratic function of diffraction angle for small angles, θ, (equivalent to the rf frequency) measured from the normal to the two parallel acoustic planes; cell separation distance, d; average wavelength, λ; and, assuming no air gap between Bragg cells, is given by

$$E(\theta) = 2[1/\cos(\theta) - 1]/\lambda \approx \pi d\theta^2/\lambda. \qquad (3.8)$$

If a dual-beam Bragg cell could be made with zero separation, the phase error would be zero, but it cannot.

In order to improve correlator performance beyond what can be achieved with uncorrected proximity imaging, it is necessary to reduce this as well as other frequency-dependent phase errors. A potential solution is to use a properly designed analog electrical network to generate a complementary frequency-dependent phase distortion function in the signal to the first Bragg cell. The equivalent would be to generate an identical distortion in the signal to the second Bragg cell. Whichever is used, the two complementary distortion functions would then cancel optically. If this could be carried out successfully it might be called 'phase-corrected proximity imaging', or 'holographic imaging', as it was called by P. Tamura, who first suggested it.

Phase/amplitude error in a conventional imaging correlator

The straightforward optical solution, which was shown in Figure 3.8, is to image one acoustic beam or optical wavefront, into the other, using conventional Fourier imaging, i.e., with an FRL having significantly less phase error than is caused by proximity imaging. In the imaging correlator, phase error results mainly from space- and frequency-dependent optical wavefront distortion in the Fourier relay and AO cells, and from electrical frequency-dependent error in the AO cell transducers and impedance matching circuits. In this type of correlator the reference beam, necessary for heterodyning, passes through the same lenses as the diffracted light, resulting in more stability and compactness than if the reference beam followed a separate path, but with less stability and compactness than would be possible with corrected proximity imaging. However, the trade-off is that the required angular working range for the lens is increased to nearly $\pm 4.5°$ for AO cells of shear mode TeO_2 operating at near-maximum bandwidth.

When an FRL is used to image diffracted light from an encoded signal on an acoustic carrier, from one AO cell to the other, good (auto)correlation requires that resolution be sufficiently high and geometrical distortion sufficiently low that the bits, or 'chips', of the reproduced, or transmitted, wavefront closely match those generated by the second Bragg cell in size and position. Also, the WFD over the back focal plane must be low enough that the relative phases of the reproduced bits match those of the second Bragg cell, except for being reversed in sense, i.e., advanced instead of retarded (complex conjugate requirement). When these conditions are met, the transmitted wavefront, after being diffracted a second time, will be nearly a plane wavefront (at the moment of peak autocorrelation), and will heterodyne properly with the reference wave. It should be mentioned that any WFD common to both waves is of little concern because it cancels in the heterodyne process.

Frequency-dependent error

The electrical frequency-dependent phase/amplitude errors in the AO transducers can be much larger than the equivalent optical error in a good FRL. However, the latter can, in principle, be made to cancel as follows: since phase error is additive in the two successive diffraction processes, use of 'complementary' phase error vs frequency curves (anti-tracking) for the two AO cells results in cancellation. In a Mach–Zehnder-type processor in which the diffracted light from the two AO cells follows separate but equivalent paths, the phase vs frequency error curves of the two AO cells would need to be identical in order to cancel. Tracking of error curves is discussed further in Section 3.5.

Aperture position-dependent error

If the reference beam path did not pass through the FRL, as it actually does, it could have an arbitrary space-dependent WFD, and a simulation of this unrealistic but instructive case, in Section 3.5, shows that a smoothly changing peak-to-valley WFD over all the heterodyned plane wave components of $\lambda/7$ results in a 3 dB correlator signal loss. This is equivalent to $\lambda/30$ rms and is well beyond the capability of conventional lenses. If the error is not smoothly varying, it may contain large amplitude sinusoidal components at a particular frequency (ripple). Ripple effects were discussed in Section 3.4.1. A phase ripple component is identical to Fourier components of surface irregularity which produce scattered light, and these were also briefly discussed in Section 3.4.1. Amplitude ripple is more likely to be produced by effects in the Bragg cells and transducers than in the FRL. The frequency response curve of some shear TeO_2 Bragg cells has a double peak, i.e., two cycles of ripple. This can sometimes be reduced by careful adjustment of alignment and collimation. Multi-pole transducer matching networks could have electrical phase ripple; however, matching circuits normally do not have multiple poles. Both need to be avoided.

Since the data arrays in the correlator example are, in general, complex, i.e., encoded in both amplitude and phase, both amplitude and phase errors generated by the correlator transducers and optics must be sufficiently small in order not to degrade the correlation operation unduly. Amplitude and phase errors between input and output planes of a Fourier lens are thus two of the important measures of basic optical performance for optical processing applications, but in a correlator it is really the accuracy, sensitivity, and dynamic range of the correlation operation that is of interest. The latter must be made directly relatable to optical quality and the specifications of the Fourier lenses. Of the three, dynamic range (peak-to-sidelobe ratio) and sensitivity (signal level) can be taken as dependent variables vs the accuracy of the correlation signal. If accuracy were sufficiently high it would be possible to monitor the phase of the correlation signal in an operational correlator to give additional information about conditions affecting the phase of the signal. However, it is much more convenient to perform a simulation to calculate the effect of given optical performance on phase of the correlator signal.

3.4.3 Fabrication constraints

Fabrication accuracy

After an optical processor is designed, taking diffraction and stray light considerations into account, and the design is shown to meet all performance requirements, it is still possible that the design can never be fabricated so that the performance approaches the design goal. This can happen if the designer does not take into account the realities of fabrication limitations and error during the design. These include mechanical accuracy limitations on surface fabrication and positioning, and uniformity of optical properties such as index of refraction and coating transmission. For example, if the design requires the positioning, centering, shape, index, or other accuracy for any of the elements or surfaces in the lens to exceed the fabricator's ability, the lens may not work and the design may have to be changed. 'Tolerancing' permits fabrication constraints to be taken into account in the design process, avoiding this problem. It is listed as 'automatic tolerancing' in Table 3.1. In the CODE V package, tolerancing can be run on any system that can be entered to check at least four different performance areas – rms wavefront, MTF, distortion, and third order aberrations.

Effect of tolerancing on fabrication

Automatic tolerancing works as follows: every variable in the design is individually perturbed, and the effect on performance calculated, giving a sensitivity coefficient for each of the variables. Then, assuming a random distribution of error direction, and using realistic fabrication tolerances, the CAD package can sum the effect of all errors and calculate the probability that lens performance will be satisfactory. If the result is, say, 0.33, it would require the fabrication of three devices in order to have a high probability that one would perform satisfactorily. If the above is not acceptable, the designer can make further changes to the design to reduce sensitivity to the offending variables and increase the probability of good performance. The goal is to make all sensitivities as small as possible, with no single one large and troublesome. Experienced lens designers usually will not initially design a system with serious sensitivity problems unless the design is unusual or complex, because they know from experience what conditions to avoid. In any case, a tolerancing analysis is good insurance.

When fabrication tolerances have been taken into account, the expected deviations in values of curvatures, glass index, etc. in the fabricated lens should not cause the performance to be unacceptable.

However, that performance which would be lost can usually be at least partly recovered by 'recomputes' of the design. The first recompute can be done after the actual indices of the glass that will be used are measured, and it may involve slight changes in curvatures and/or spacing. This is necessary because indices of the actual glass will differ from the handbook values used in the original design. A final recompute can be done after the lens elements have been fabricated and measured, which, of course, can only involve adjustment of lens element spacing. Another method for increasing the probability of getting good performance can sometimes be used. It involves fabricating extra components and carefully choosing and matching them during assembly using interferometers or other test methods.

Estimating fabrication feasibility

Before a design is attempted, it is desirable to be able to estimate the fabrication accuracy that would be needed for a given wavefront accuracy. If the approximate number of surfaces to be used in the design is known, and assuming that all the error is due to surface inaccuracy, that all surfaces contribute equally, and that the error direction is random, then the error, S, that each surface can contribute is

$$S = (T^2/n)^{1/2}, \tag{3.9}$$

where T is the total error and n is the number of surfaces contributing to the error. Conversely, if the ability of a fabricator to produce a surface with error S is known, the total probable wavefront performance for a given number of surfaces can be estimated. For example, the requirement for the correlator FRL is $\lambda/7$ or 0.143 waves peak-to-valley. For 26 surfaces (24 glass and 2 TeO$_2$ surfaces), S is $\lambda/36$ or 0.0278 waves peak-to-valley at 830 nm. If measurements are made using 633 nm light, this reduces to $\lambda/27$. The surface tolerance will actually be less severe than indicated because some surfaces are used at less than full aperture, which tends to reduce their error contribution. On the other hand, there will be additional wavefront error due to glass inhomogeneity, decentering and tilt of elements, etc., so that this is a best-case estimate. When tolerancing calculation of the actual design is used and all factors are taken into account, the required surface tolerance is $\lambda/20$ using 546 nm light, or $\lambda/23$ using 633 nm light. Such accuracy can be handled by a limited number of fabricators and so the design is feasible.

In summary, basic optical performance, as determined by optical CAD package ray trace plus FFT (point spread function), or sometimes by estimating procedures, can be used to calculate directly processor performance due to most physical optical constraints except scattered light: e.g., crosstalk and dynamic range of the rf spectrum analyzer for simple cw (unmodulated) signals. If BTDF data are available, a quick estimate of scatter may be possible in simple cases. If a special optical CAD package capable of modeling scattered light is also available, the effect of scatter in more complex systems can be calculated. Sometimes further modeling can be carried out using standard CAD packages, e.g., further modeling for determining the effect of other signal conditions, such as variable pulse length conditions in the rf spectrum analyzer, which is described in Section 3.5.

For more complicated processors optical CAD packages alone may not be sufficient to predict dynamic range, sensitivity, or other factors. Additional required information includes a model of the exact operations to be performed by the processor and the nature of the signals to be processed. An example of this type of modeling is also found in Section 3.5 – for the 1-D AO space integrating correlator. The exact operations to be performed are contained in a computational model of the processor, which is programmed in one of the mathematical simulation packages mentioned in Section 3.3. Inputs to the model include: (1) the basic optical performance obtained from optical CAD packages such as phase and/or amplitude distortion), (2) test signals with the desired information content and modulation method, and (3) any additional phase and amplitude distortion due to electronics.

3.5 Application of CAD tools to specific cases

3.5.1 1-D AO space integrating correlator

We will now discuss methods of simulating the operation of the imaging 1-D AO space integrating correlator, whose operation was described in Section 3.4 and diagrammed in Figure 3.8. It was mentioned that, in this case, everything between the acoustic planes of the Bragg cells, primarily the FRL and inner half of each Bragg cell, should be considered a grating interferometer, and it is the part requiring the most attention in wavefront/phase performance analysis. It was also mentioned that the electrical phase/amplitude distortion due to Bragg cell transducers is also important. This example primarily looks at phase distortion and its effect on sensitivity, or correlation signal level, and peak-to-sidelobe ratios. Other important performance factors – amplitude error, geometrical distortion and magnification error – can be included in the model, but will not be in the following.

Obtaining FRL 'wavefront performance for correlation'

In a real lens, since each plane wave component travels in a different direction through a slightly different part of the lens, each component can have a slightly different WFD, direction, and time delay. In agreement with the properties of a Fourier relay, outlined in Section 3.4.2, the wavefront performance of the lens design can be analyzed as follows: the plane wave performance (infinite conjugate case) of a design can be obtained by passing plane waves (rays) through the input pupil at the front focal plane, at desired angles over the operating range, and computing the relative WFD using the OPD routine in a standard CAD package, over each plane wave at the exit pupil or back focal plane. These OPD plots do not retain absolute phase information because the 'piston', or average optical path, has been subtracted. It is necessary to reinsert the piston, which is removed by standard CAD packages, in order to have absolute phase, or time of arrival, informa-tion. This can be done by calculating one more OPD, that of a single point centered in the input pupil. This is a ray 'fan' over the design angular range. Rays in the fan corresponding to the directions of the plane wave components can be seen to be the 'chief', or 'principal' rays of the plane wave components. Non-zero OPDs for the chief rays is equivalent to frequency-dependent phase error, or error in time of arrival. If the lens has very good on-axis imaging, the OPDs of the chief rays will be small. Any significant OPDs of the chief rays must be summed (subtracted or added) with the OPD plots of all the plane waves as a correction factor in order that they are referenced to each other in phase.

The next step in modeling correlation wavefront performance is to simulate diffraction at the second Bragg cell (combining beam splitter) and mixing of all components with the reference wavefront (hetero-dyning). This is done by subtracting the corrected OPD curve of the chosen (undiffracted) reference beam from all the corrected OPD curves of the diffracted components, since this is what occurs when the transmitted wavefront is diffracted and heterodyned with the reference wavefront. When the plane wave WFD of a symmetrical 1:1 relay lens has mirror symmetry about the center of the beam, it is not necessary to 'flip', or reverse, the WFD curve of the reference 180° before

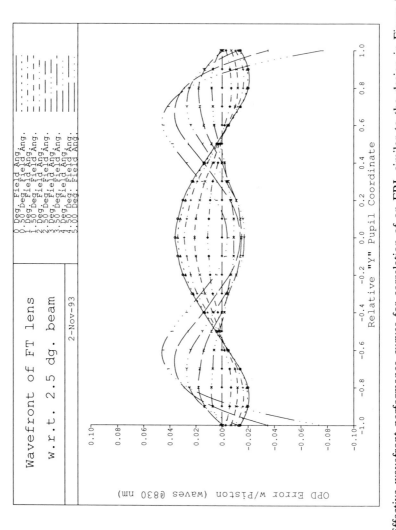

Figure 3.10 Effective wavefront performance curves for correlation of an FRL similar to the design in Figure 3.6, except that Bragg cells are included in the design, and curves are calculated with the Bragg cells in place.

subtracting, even though the reference beam is on the opposite side of the optical axis and would otherwise need to be reversed. For the case of a relay system without the above symmetry, the reversal is necessary.

It should be mentioned that the manipulation of OPD data just described is not a built-in feature of any known CAD package, but can be done using the 'post-processing' capability in certain CAD packages, such as the macro capability in CODE V. It should also be mentioned that the above procedure could be followed to *measure* the wavefront performance for the correlation of a fabricated lens using a standard Fizeau phase shift interferometer to generate both the sequence of plane waves and the imaging wave. To obtain the correlator performance of the lens, the same correction and subtraction of wavefronts as above are needed. Some computer-controlled interferometers permit wavefronts to be stored in memory and manipulated, which facilitates the measurement.

FRL wavefront performance for correlation: actual FRL design example

The FRL was specified to have an angular working range of $\pm 5°$ with no clipping, an effective wavefront distortion for correlation, as described above, of $\lambda/7$ or better, magnification of 1 ± 0.00025 and equivalently low geometrical distortion. A layout at one stage in the design process of this lens was shown in Figure 3.6. 1-D plots of corrected and heterodyned wavefronts, as described above, are given in Figure 3.10, at $\frac{1}{2}°$ intervals out to 5°. It can be seen that the WFD is slightly different in shape for each direction over the angular range of operation, and that the absolute peak-to-valley wavefront distortion is 0.06λ ($\lambda/17$) for most plane wave components over most of the aperture. The 5 degree component is much worse, 0.12λ ($\lambda/8$). This is a very high angular working range for such a lens, and since the desired performance of the finished lens is $\lambda/7$, and the design performance is considerably higher, the chances of achieving the desired working performance are high, although the lens system has not been fabricated at the time of writing.

Bragg cell considerations

The constraint of including the Bragg cell optical material in the FRL design was discussed in Section 3.4, and WFD performance curves for the case of Bragg cells included in the design are shown in Figure 3.10. When they are not included in the design but are included for the

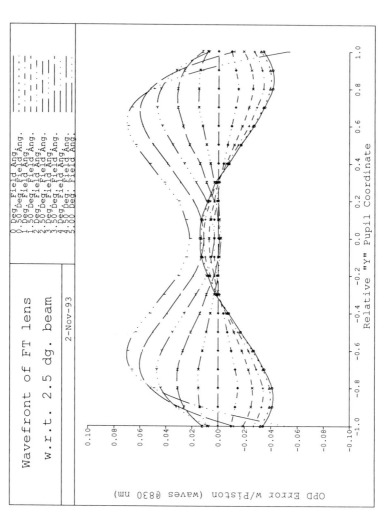

Figure 3.11 Effective wavefront performance curves for correlation of an FRL similar to the design in Figure 3.6, except that Bragg cells are not included in the design, but are inserted for the purpose of calculating the curves.

purpose of calculating performance, or vice versa, the peak-to-valley WFD is increased by about a factor of 2. This case is shown in Figure 3.11. Some of the errors due to the basic Bragg cell diffraction process (Hecht, 1977), i.e., depletion, cross modulation, IMD etc., do have optical design implications but are not considered in this chapter. However, it should be mentioned that intrinsic third order IMD due to two or more simultaneous signals in an AO device is analogous to IMD in electronic amplifiers, and must be considered and controlled so that it does not become the limiting factor in the optical processor.

The Bragg cell rf impedance matching network and the acoustic transducer itself can contribute significant frequency-dependent phase and amplitude error over the operating band and will be considered next. If there were no electrical or acoustical frequency-dependent phase error, phase would be a linear function of frequency in the rf impedance matching network and the Bragg cell acoustic transducers. However, it is very difficult to fabricate Bragg cells with low deviation from linear phase. Also, in this correlator, deviations from linearity in the two Bragg cells add because it is an optical multiplication process. Therefore, if the Bragg cells can be fabricated so that one phase error function is approximately the complement of the other (anti-tracking), their deviations from linearity would tend to cancel. (A similar approach was mentioned earlier for the proximity imaging correlator.) The part that does not cancel would sum with the small optical phase vs frequency error function (chief ray OPD) of the lens.

Modeling correlator performance: frequency-dependent phase error only
An example of the calculated effect of frequency-dependent phase error on correlator performance will now be given using the Mathematica package for various assumed phase error functions. For the present simulation, MTF-related amplitude error and geometrical distortion error will be assumed negligible. Fourier-transform-based correlation simulation is used (see Eq. (3.6)), since frequency-dependent error is best introduced in the frequency domain, and this provides easy access to the frequency domain. The basic correlation operation used in the error model is given by Eq. (3.10).

$$\text{CrossCor(list1, list2)} = \text{IFFT}[\text{FFT(list1)}*f_{\text{err}}*\text{Conj}\{\text{FFT(list2)}\}]$$

$$(3.10)$$

Here, f_{err} is the frequency-dependent phase/amplitude error function with amplitude assumed constant, and takes on various assumed shapes and peak-to-valley values over the frequency band. List1 and list2 are identical $\pm\pi/2$ radian binary random phase modulated data sets of 1022 nonzero values, padded with 1022 zero values to prevent wraparound errors in the calculation, hence we are modeling the autocorrelation of a random sequence. The minimum optical aperture size corresponds to 1022 data elements. It was found that the correlation signal level is not highly sensitive to the exact (smoothly varying) shape of the error function, but was sensitive to the assumed peak-to-valley value of phase error. Sensitivity will be defined here as the fall-off, in dB, of the simulated correlation signal with optical errors present, compared to a similar calculation with no errors present. Because there is no aperture position-dependent error assumed (it is treated later), there is no need to provide for the effect of fixed optical phase error on moving data, and the result for a given f_{err} is obtained quickly from a single cycle of the above operation. Although this is a fairly simple calculation, care is needed to ensure that the frequency terms of the error function multiply the corresponding frequency terms of the signal data.

The results in Table 3.2 show that when the peak-to-valley phase error approaches $2\pi/5$, the correlation signal fall-off is about 4 dB, and it is not very sensitive to the exact shape of the error function, i.e., half cosine, full sine, or linear (not shown). When it is $2\pi/5.45$ rad, or 66°, the correlation sensitivity is down about 3 dB. This value is of interest because it is believed that the impedance of two TeO_2 Bragg cell transducers can be made to track this well (private communication, E. Young, Newport EOS, Melbourne, FL, 1993). Also, it is of interest because this error can result from the transducers as well as from the

Table 3.2 *Correlator sensitivity and peak/sidelobe (rms) vs frequency-dependent phase error.*

	Half cosine		Full sine	
Phase error (rad)	Sensitivity (db)	Peak/sidelobe ratio (rms)	Sensitivity (dB)	Peak/sidelobe ratio (rms)
$2\pi/10$	0.8	31.4	0.9	31.3
$2\pi/7$	1.8	29.9	1.9	29.8
$2\pi/5.45$	3.1	28.1	3.3	27.8
$2\pi/5$	3.8	27.2	4.0	26.9

relay lens, as discussed previously, i.e., in principle, it would be possible to use transducer error to cancel frequency-dependent optical error in the relay lens because these errors sum in this architecture; however, considerable error would have to be designed into the lens, which would not be necessary if a 3 dB signal loss due to Bragg cells were acceptable.

Modeling correlator performance: aperture position-dependent phase error only

An example of the calculated effect of an uncompensated aperture position-dependent phase error on correlator performance will now be given. This simple example applies more realistically to an imaginary 'Mach–Zehnder'-type correlator shown in Figure 3.9, in which the reference beam follows a separate path which can have an uncompensated relative WFD, or OPD, with respect to the diffracted path. The uncompensated WFD must be assumed to be the same for all Fourier components. The smoothly-varying aperture position-dependent phase error function, for which data are given, was a half-cycle sinusoid, but again the peak-to-valley error and not the shape of the error function was found to be most important.

This case was modeled using the shift/multiply/sum method of correlation (see Eq. (3.7)), which permits the signal data to shift with respect to the fixed optical error, but does not permit the introduction of frequency-dependent phase error, or any variation of aperture-dependent error for different plane wave components. In this correlation operation the single error function is fixed and the data are shifted in opposite directions (as in the optical correlator) over the index m, giving one value of the correlation function for each value of m.

$$\text{CrossCor(list1, list2)}(m) = \sum_{n} \text{list1}(-n) * \text{error(fixed)} * \text{list2}(n + m)$$

$$(3.11)$$

As before, list1 and list2 are identical and modulated as a $\pm \pi/2$ rad random binary phase, so that phase errors would be expected to have an increasing effect on the correlation signal as the peak-to-valley phase error approaches $\pi/2$. However, the interesting result for this case is that there is cyclic nulling when the peak-to-valley phase errors are, very roughly, multiples of a quarter wave, as shown in Figure 3.12. When peak-to-valley phase error reaches $\lambda/5$ or 1.26 rad, a typical wavefront-distortion specification for a TeO_2 Bragg cell, the

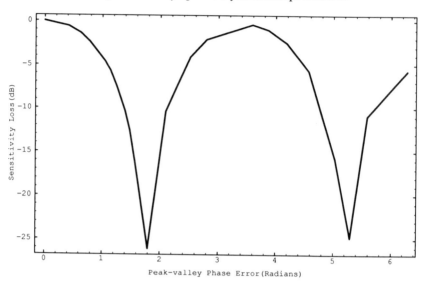

Figure 3.12 Correlation sensitivity vs aperture position-dependent phase error.

correlation signal fall-off is over 7 dB, and reaches the first null at $\lambda/3.5$ or 1.80 rad. The correlation sensitivity function for this case may be related to a sinc function.

Modeling correlator performance: actual FRL design

For this case, there is both aperture position-dependent phase error and frequency-dependent phase error. The corrected and heterodyned WFD plots for this case, Figure 3.10, were seen to change slowly with working angle. Previously, when using the multiply/shift/sum method for aperture-position-dependent error, all plane wave components had to have the same uncompensated WFD, and all components were assumed to be passed. For the present case with a slightly different WFD for each of a finite number of plane wave components, with moving data, the situation is more complicated. The transform method, Eq. (3.6), might be combined with the shift/multiply/sum model, Eq. (3.7), in order to be able to simulate the effect of moving data, as well as introduce frequency-dependent error. It is also necessary to be able to introduce slightly different WFD data for each Fourier component that is passed by the lens in order to have an accurate simulation. Using a package like Mathematica with the FFT and IFFT transform routines is very inefficient since almost all of the calculated output must be thrown away, i.e., the correlation must be calculated for each position of the moving data with respect to the unique error function for each Fourier component, and only one data point corresponding to each position of the moving data saved. This might be done in a reasonable time if a fast routine were written in a compiled language using a library like Prime Factor FFT, but at the time of writing this has not been done. However, for an FRL with small frequency-dependent phase error so that piston correction is not necessary, and considering the phase error to be in the total integrated (diffracted) wavefront with respect to the reference wavefront, the previous cyclic nulling result for a 'Mach–Zehnder' correlator can be considered to be an approximation for the present case. The above approximation will be more realistic for narrow bandwidth operation when the diffracted waves cover a relatively small angular range.

3.5.2 1-D AO spectrum analyzer

The 1-D AO spectrum analyzer was discussed in Section 3.4.1. Here we look at how standard CAD package capability can be used to estimate basic diffraction performance. Table 3.3 shows how this and other factors combine to give total crosstalk and dynamic range.

Point spread function with detector integration

The point spread function (PSF) is a basic and powerful analysis tool for assessing processor design performance, and is available in several standard packages. It permits a diffraction-based imaging estimate of a single transform lens or group of lenses for a particular collimated input illumination. When used with detector integration, such as is available in CODE V, it permits the assessment of light distribution across a particular detector array geometry, or of signal levels due to crosstalk. The actual laser source and its optics may not be well known at this point, or it may be desired to vary experimentally the shape (apodization) and/or size of the laser beam. For example, the simplest form of AO rf spectrum analyzer consists of a single wideband Bragg cell illuminated by a collimated laser beam, a Fourier transform lens, and a photodetector array where each pixel represents a frequency bin. The signal of interest may consist of rf pulses which are either longer or shorter than the illuminated region of the Bragg cell. Those which are longer are seen as cw signals, and with optimum laser beam apodization will have the most compact frequency channels (spot size) in the detector plane.

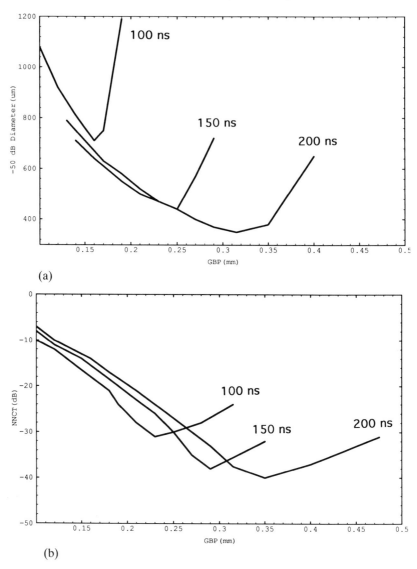

Figure 3.13 PSF calculation of rf spectrum analyzer performance for 100, 150, and 200 ns rf pulse lengths and Gaussian illumination: (a) −50 dB spot diameter; (b) NNCT.

An example of a simple use of the PSF tool is choosing the optimum laser beam apodization for the cw signal case. For uniform intensity across the laser beam (no apodization) the spot size at the detector plane is broadened by sidelobes of the sinc function, and for Gaussian apodization the sidelobes decrease and the central spot size broadens as the Gaussian beam parameter (GBP), or beam size, is reduced. For a given detector geometry, the optimum apodization is one which gives the greatest optical dynamic range at the photodetector, i.e., the greatest ratio of signal on a pixel to signal on adjacent pixels. Hence if the GBP is 'zoomed' over a range of values while running PSF with integration over detector geometry, the optimum Gaussian apodization can be quickly located. An example of a package which permits simple detector geometry to be specified and provides automatic integration of PSF over the detector is CODE V.

An example of a more difficult analysis goal is when there is a range of signal types or pulse lengths of interest, a choice of detector pixel size and spacing, and a maximum size the processor can occupy. A selected design can be tested by repeating the previous PSF calculation for a number of pulse lengths over the range of interest, and choosing the apodization which gives the best or acceptable overall performance. If performance is still not acceptable the design must be altered. One result of a series of PSF calculations for a particular rf channelizer design is shown in Figure 3.13 for three different rf pulse lengths (the equivalent of three different limiting optical apertures) vs GBP. Nearest neighbor crosstalk (NNCT) is for a particular detector array geometry with 10 mil centers for 7 by 9 mil pixels. It can be seen that, for this particular processor, NNCT for the longest pulse of interest can, at best, be 40 dB below the signal for a GBP of 0.315 mm, and that shorter pulse lengths will have higher crosstalk. A value of GBP can then be chosen which may give an acceptable NNCT for all three pulse widths. The PSF calculation will be useful whenever the 'impulse response' of a lens or group of lenses to a particular collimated beam is needed. It will usually be used early in the design process before the analysis of stray and scattered light is undertaken, but in the final analysis the combined result of all factors must be summed.

In a spectrum analyzer of the type described above, a number of contributing factors combine to determine the total performance. In addition to the basic diffraction performance, there is stray light, IMD, birefringence, and detector optical and electrical crosstalk performance. 'Birefringence' refers to the fact that some Bragg cell designs have an unwanted low-level mode (for light of a different polarization) offset in frequency from the desired mode. In Table 3.3, measured or estimated values for all factors are assumed to sum to give a worse-case estimated total NNCT performance. For some cases an rms or other

Table 3.3 *NNCT contribution of various determining factors (dB) for an rf spectrum analyzer/channelizer.*

Factor	Measured/calculated	Goal
Diffraction	−28	−50
Stray light	−30	−50
IMD, Bragg cell	−30	−50
Birefringence, Bragg cell	−33	−50
Detector crosstalk	−35	−50
Total, all factors	−24	−43

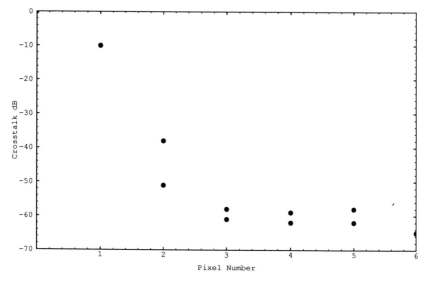

Figure 3.14 Optical crosstalk/sidelobe performance comparison of two different doublet Fourier transform lenses. The higher values are for a cemented doublet, the lower values are for an air-spaced doublet. The lenses had the same diameter, focal length, Gaussian illumination, and pixel geometry.

sum may be more appropriate. These values were determined for a particular set of available components.

PSF can also be used to compare the diffraction performance of slightly different optics for the same conditions. In Figure 3.14 we compare the crosstalk/sidelobe performance of two different Fourier transform lenses when illuminated by an on-axis collimated Gaussian beam, and with a particular detector array in the focal plane. PSF is integrated over pixel geometry. The higher values are for a simple cemented doublet lens and the lower values are for an air-spaced doublet. There is no difference in performance for the nearest neighbor pixel.

3.5.3 Related tools

Modulation transfer function

The MTF calculation is a very important basic tool for determining the optical frequency response of a lens design. It is the counterpart of an MTF measuring system for measuring a finished lens. The calculation can be geometric(ray)- or diffraction-based. Sometimes both are available in the same CAD package. If so, this provides additional information for the designer, as does comparing the geometrical ray trace spot size with the equivalent PSF spot size. If the geometrical spot is smaller than the diffraction-based spot, it is likely that sufficient aberration correction has been performed on the system and the system is diffraction-limited. Likewise, if the geometric MTF is better than the diffraction MTF, the system is diffraction-limited and the geometric MTF is not used. Also the MTF can be computed for square wave as well as sine wave response in some packages.

A diffraction-based calculation of the MTF for the version of the FRL shown in Figure 3.6 is given in Figure 3.15. It shows the level of response at the output plane due to an input beam with sinusoidal amplitude variation. The upper dotted curve is the calculated diffraction-limited, or perfect, response, and the others are for 0, 3.5, and 5° fields and for radial and tangential directions in the image. There is little separation between the curves, which shows that the MTF of the lens is very nearly diffraction-limited and changes very little with field angle. The MTF (amplitude) data for the FRL may be included in the mathematical model along with other amplitude distortion data, the phase distortion data, geometrical distortion data, and modulated signal data, to give improved performance estimates. However, the amplitude was assumed constant in the 1-D AO correlator examples.

Partial coherence analysis

Another related analysis tool, 'partial coherence analysis', called partial coherence (PAR) in CODE V, is useful for calculating the diffraction response of a lens to a specific input test pattern such as a three-bar or a five-bar intensity pattern. In CODE V the calculation

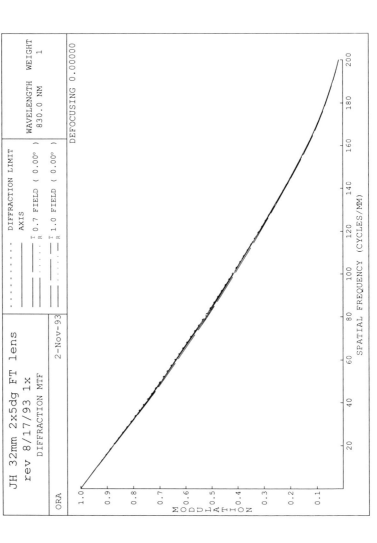

Figure 3.15 Sine wave diffraction MTF for the FRL of Figure 3.6, for 830 nm light. (Optical Research Associates, Pasadena, CA.)

can be performed for different levels of coherence and combinations of wavelengths, hence the name. Calculations for a five-bar input intensity pattern are shown in Figure 3.16 for 830 nm light and for the two extremes of 100% coherence and 0% coherence. Such plots are a composite of a number of MTF-type calculations. The response of lenses to intensity patterns is more useful in modeling processors which use SLMs than in modeling AO processors, in which the modulation choice is usually phase. An optical CAD package with phase object capability is not known to the authors at the time of writing. Hence, in order to calculate phase objects such as AO modulated data, one of the mathematical simulation packages must be used for this purpose, transferring phase and amplitude error data for the lens from the optical CAD package to the simulation package and inputting the phase object. This, of course, was the method used in the 1-D AO space integrating correlator examples.

3.6 Conclusions

In summary, basic optical performance as determined by optical CAD package ray trace plus FFT (PSF), or sometimes by estimating procedures, can be used to calculate directly processor performance resulting from most physical optical constraints with the exception of scattered light: for example, crosstalk and dynamic range of the rf spectrum analyzer for simple cw signals. If BTDF data are available, a quick estimate of the effect of scatter may also be possible in simple cases. If a special optical CAD package capable of modeling scattered light is also available, the effect of scatter in more complex systems can be calculated. Sometimes modeling, in addition to basic optical performance, can be carried out using standard CAD packages: e.g., repeated application of the PSF and detector integration to map system performance for variable pulse length conditions in the 1-D AO rf spectrum analyzer example.

For more complicated processors, optical CAD packages alone may not be sufficient to predict dynamic range, sensitivity, or other factors. Additional required information includes a model of the exact operations to be performed by the processor and the nature of the signals to be processed. An example of this type of modeling is also found in Section 3.5 for the 1-D AO space integrating correlator. The exact operations to be performed are contained in a computational model of the processor, which is programmed in one of the mathematical

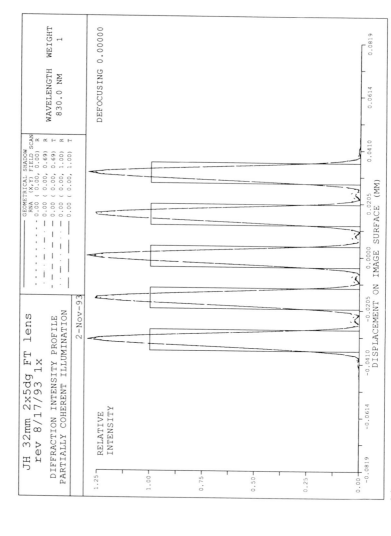

(a)

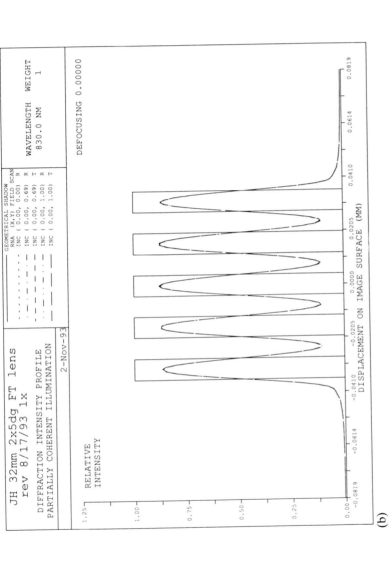

Figure 3.16 Diffraction intensity response for the FRL of Figure 3.6 to the five-bar intensity pattern with a 16 μm period, or 62.5 cycles/mm fundamental. Input pattern is shown superimposed on output: (a) 100% coherent illumination; (b) 100% incoherent illumination (Optical Research Associates, Pasadena, CA).

simulation packages mentioned in Section 3.3. Inputs to the model include: (1) the basic optical performance obtained from optical CAD packages such as phase and/or amplitude distortion, (2) test signals with the desired information content and modulation method, and (3) any additional phase and amplitude distortion due to electronics.

There is still a considerable gap between what is needed to design and simulate coherent optical processors and what is available to do the job. No single tool at present is sufficiently powerful for that purpose. Standard optical CAD packages, as developed by lens designers for lens design, are continually becoming more powerful but still lack some features desirable for coherent optical processor design. However, the gap is narrowing with continued improvements to standard CAD packages, with the introduction of special optical CAD packages (which allow quick simulation of scatter and multiple reflections in a processor design), and with the continued development of powerful mathematical packages which can quickly do signal processing calculations for system simulation. But there is still not a *single* comprehensive CAD tool available for optical signal processing. Also, it is not likely that the additional features needed will be added to any major CAD package in the near future to produce such a comprehensive tool, due to the relatively small demand, and the need to restrict the features of major packages, such as CODE V, to those with the greatest utility to all areas of optical design.

References

Anderson, G. W., Guenther, B. D., Hynecek, J. A., Keyes, R. J., and VanderLugt, A., Role of Photodetectors in Optical Signal Processing. *Applied Optics*, **27**, 2871 (1988).

Anderson, G. W., Webb, D., Spezio, A. E., and Lee, J. N., Advanced Channelization Technology for RF, Microwave, and Millimeterwave Applications. *Proc. IEEE*, **79**, 355–88 (1991).

Bennett, H. E., Scattering Characteristics of Optical Materials. *Optical Engineering*, **17**, 480–8 (1978).

Berg, N. J., Lee, J. N., Casseday, M. W., and Katzen, E., Adaptive Fourier Transformation by Using Acoustooptic Convolution. In *Proc. 1978 IEEE Ultrasonics Symp.* IEEE No. 78CH1344-1, ed. J. deKlerk & B. R. McAvoy, pp. 91–9. New York: Institute of Electrical and Electronics Engineering (1978).

Blandford, B. A. F., A New Lens System for use in Optical Data Processing. In *Optical Instruments and Techniques 1969*, ed. J. H. Dickson, pp. 435–43. Newcastle upon Tyne: Oriel Press (1970).

Brown, R. B., Craig, A. E., and Lee, J. N., Predictions of Stray Light

Modeling on the Ultimate Performance of Acousto-optic Processors. *Optical Engineering*, **28**, 1299–305 (1989).

Buralli, D. A., and Morris, G. M., Design of a Wide Field Diffractive Landscape Lens. *Applied Optics*, **28**, 18 (1989).

Casasent, D., Optical Information-Processing Applications of Acousto-Optics. In *Acousto-optic Signal Processing: Theory and Implementation*, N. J. Berg and J. N. Lee, eds., p. 331. New York: Marcel-Dekker (1983).

Chang, I. C., and Lee, S., Efficient Wideband Acousto-optic Cells, *Proc. IEEE Ultrasonics Symp.*, IEEE No. 83CH 1947-1, ed. B. R. McAvoy, p. 427. New York: Institute of Electrical and Electronics Engineers (1983).

Church, E. L., and Zavada, J. M., Residual Surface Roughness of Diamond-turned Optics. *Applied Optics*, **14**, 1788–95 (1975).

Dereniak, E. L., Brod, L. G., and Hubbs, J. E., Bidirectional Transmittance Function Measurements on ZnSe, *Applied Optics*, **21**, 4421 (1982).

Farnett, E. C., Howard, T. B., and Stevens, G. H., Pulse-compression Radar. In *Radar Handbook*, M. I. Skolnik, ed., Chapt. 20, New York: McGraw-Hill (1970).

Faklis, D., and Morris, G. M., Optical Design With Diffractive Lenses (Part I), *Photonics Spectra*, November, 205–8 (1991).

Forkner, J. F., and Kuntz, D. W., Characteristics of Efficient Laser Diode Collimators. *Proc. SPIE*, **390**, 156 (1983).

Goodman, J. W., Introduction to Fourier Optics. New York: McGraw-Hill, (1968).

Goutzoulis, A. P., Gottleib, M., and Singh, N. B., High Performance Acousto-optic Materials: Hg_2Cl_2 and $PbBr_2$. In *Proceedings of the SPIE*, Vol. 1704, ed. Pape, D., pp. 195–209. Bellingham, WA, USA: SPIE (1992).

Griffin, R. D., and Lee, J. N., Acousto-optic Wide Band Correlator System Design, Implementation, and Evaluation, *Applied Optics* **33**, 6774 (1994).

Harris, F. J., On the Use of Windows for Harmonic Analysis with the Discrete Fourier Transform, *Proc. IEEE*, **66**, 51 (1978).

Hecht, D. L., Multifrequency Acoustooptic Diffraction, *IEEE Trans. Sonics and Ultrasonics*, **SU-24**, 7–18 (1977).

Lee, J. N., Optical Architectures for Temporal Signal Processing. In *Optical Signal Processing*, ed. J. L. Horner, pp. 165–90. San Diego: Academic Press (1987).

Lee, J. N., and VanderLugt, A., Acousto-optic Signal Processing and Computing, *Proc. IEEE*, **77**, 1528–56 (1989).

Lee, S. H., Recent Advances in Computer Generated Hologram Applications, *Optics and Photonics News*, July, 18–23 (1990).

Nicodemus, F. E., Directional Reflectance and Emissivity of an Opaque Surface, *Applied Optics* **4**, 767 (1965).

Scifres, D. R., Lindstrom, C., Burnham, R. D., Streifer, W., and Paoli, T., Phase-locked (GaAl)As Laser Diode Emitting 2.6W CW From a Single Mirror. *Electron. Lett*, **19**, 169 (1983).

Smith, W. J., *Modern Optical Engineering*, New York: McGraw-Hill, Inc. (1966).

Spezio, A. E., Lee, J. N., and Anderson, G. W., Acousto-Optics for System Applications, *Microwave J.*, February, pp. 155–63 (1985).

Stover, J. C., Bjork, D. R., Brown, R. B., and Lee, J. N., Experimental Measurement of Very Small Angle Stray Light Optical Performance of

Selected Acousto-Optic Materials. In *Growth and Characterization of Acousto-Optic Materials*, eds. N. B. Singh and D. J. Todd, pp. 56–93. Zurich: Trans Tech Publications Ltd (1990).

Stover, J. C., *Optical Scattering Measurement and Analysis*, New York: McGraw-Hill, Inc. (1990).

Torrieri, D., Introduction to Acousto-optic Interaction Theory. In *Acousto-optic Signal Processing: Theory and Implementation*, eds N. J. Berg and J. N. Lee, pp. 17–18. New York: Marcel-Dekker (1983).

VanderLugt, A., *Optical Signal Processing*, New York: John Wiley & Sons, Inc., pp. 8 & 128 (1992).

Veldkamp, W. B., and McHugh, T. J., Binary Optics. *Scientific American*, May, 92–7 (1992).

Editor's notes:

Recently Optical Research Associates announced a forthcomming optical system modeling package called LightTools™ which will likely be an addition to the special CAD tools in Section 3.3.2.

GENII has been purchased by Sinclair Optics and is available through them.

4

Comparison between holographic and guided wave interconnects for VLSI multiprocessor systems

MICHAEL R. FELDMAN, JEAN L. CAMP,
ROHINI SHARMA and JAMES E. MORRIS
Department of Electrical Engineering, University of North Carolina

4.1 Introduction

This chapter is dedicated to the evaluation of optical interconnects between electronic processors in multiprocessor systems. Each processor is fully electronic except for the incorporation of a number of photodetectors and optical signal transmitters (e.g., laser diodes, light emitting diodes (LEDs) or optical modulators).

The motivation for this analysis stems from fundamental advantages of optics as well as those of electronics. While electrons are charged fermions, subject to strong mutual interactions and strong reactions to other charged particles, photons are neutral bosons, virtually unaffected by mutual interactions and Coulomb forces. Thus, unlike electrons, multiple beams of photons can cross paths without significant interference. This property allows holographic interconnects to achieve a 3-D (three-dimensional) connection density with only 2-D optical elements. Similarly, photons can propagate through transparent materials without appreciable attenuation or power dissipation. Thus, neglecting speed of light delays, the speed of an optical link is limited only by the switching speed and capacitance of the transmitters and detectors. (For a 10 cm connection length, and a 50° hologram deflection angle, the speed-of-light delay is 0.5 ns.) Hence, the speed and power requirements of an optical interconnect are independent of the connection length. Since electrical very large scale integration (VLSI) connections have a switching energy directly proportional to the line length and an RC delay that grows quadratically with line length, for long enough communication links, optical connections will dissipate less power and provide faster data rate communication. Thus photons are particularly well suited to communication and interconnects, while electrons are particularly well suited to switching and logic functions.

Therefore, it is natural to conclude that a high performance computer can be made from electronic logic and processing functions connected by optical communication links. This is equivalent to taking a conventional fully electronic computer and replacing some of the electronic 'wires' with optical interconnects. (By 'wires' we mean all electronic interconnects within a computer including coaxial cable, PC board lines, multi-chip module lines, and metal lines integrated onto an integrated circuit.)

The question remains as to what percentage or which 'wires' should be replaced with optical links. While some researchers have proposed that all wires should be replaced, most agree that only the longer interconnects should be replaced. The argument behind this reasoning is as follows: (1) There is a power loss associated with conversion between optical and electrical signals through the optoelectronic devices and associated circuitry. If this power loss negates the power gained through the advantages of optical propagation, there is little or no advantage in using optical interconnects for these cases. (2) The cost of optical interconnects is much higher than that of electrical interconnects.

However, if optical interconnects are used only to replace particular electrical connection links the following advantages can be attained: (1) increased communication speed, (2) reduced power consumption, (3) reduced area, (4) reduced crosstalk, (5) increased reliability, (6) increased fault tolerance, (7) reduced cost, (8) enhanced circuit testing capabilities, (9) provision of a path toward dynamic interconnects.

These improvements may be achieved with either free space or guided wave optical interconnects, depending on the technological issues involved in the specific implementation. Many manufacturers are currently trying to decide whether to invest resources into guided wave or free space optical interconnects. On one hand, free space interconnects, due to their 3-D nature, have fundamental advantages over guided wave interconnect systems (Feldman & Guest, 1987). On the other hand, guided wave systems have advantages in terms of packaging, simplicity of fabrication and possibly cost (Guha, Briston, Sullivan & Husain, 1990). Therefore it is important to quantify the specific performance trade-offs.

In particular in this chapter guided wave interconnects, based on rather optimistic assumptions, are compared to realistic present-technology based models of free space interconnects. The point of this comparison is to illustrate the minimum performance advantages that can be expected from free space optical interconnects despite potential advances in the technology of guided wave interconnect systems.

The particular free space optical interconnect system to be considered here consists of semiconductor laser diodes, photodetectors and holographic optical elements (Figure 4.1). The holograms are used to connect the lasers and detectors in the desired pattern (Feldman & Guest, 1987). Two free space optical interconnect systems are considered: (1) the space-variant, double pass holographic system (Feldman & Guest, 1989) and (2) the basis set system (Jenkins, Chavel, Forchheimer, Sawchuck & Strand, 1984; Stirk, Athale & Haney, 1988; Thompson, 1979). These two systems are illustrated in Figures 4.1 and 4.2.

In the double pass system, the hologram is divided into many subholograms. 'Source subholograms' are placed over each laser and 'detector subholograms' are placed over each detector. Each source subhologram divides the incident wavefront into F beams (for a fanout of F), deflects each beam at the appropriate angle and changes the divergence angles of the beam so that after reflection from the planar mirror each beam is focused onto the appropriate detector subhologram. Each detector subhologram focuses the incident beam onto the detector located directly below it.

In the basis set system, described in Jenkins *et al.* (1984), Stirk *et al.* (1988), and Thompson (1979) and illustrated in Figure 4.3, a lens is used to perform a Fourier transform of the input plane. A computer generated hologram (CGH), consisting of M facets, is placed in the Fourier plane, and a second lens is used to perform a Fourier transform between the Fourier plane and output plane. Each CGH facet in

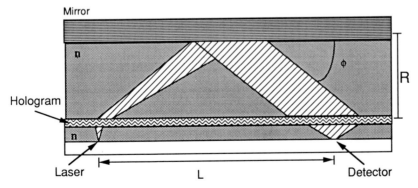

Figure 4.1 A double pass hologram system configuration as assumed in the derivations.

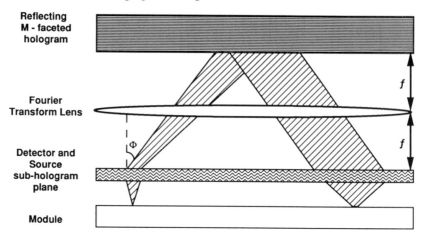

Reflecting M - faceted hologram

Fourier Transform Lens

Detector and Source sub-hologram plane

Module

Figure 4.2 The folded basis set system.

the Fourier plane implements a different connection. This requires that each connection pattern can be formed by summing together a subset of the M connection patterns implemented by the M facets in the Fourier plane.

A compact version of the basis set system, termed the folded basis set, is illustrated in Figure 4.2. The upper CGH contains the Fourier plane subholograms. Note that the planar mirror in Figure 4.1 is replaced by a reflective CGH. The lens in Figure 4.2 implements the function of both of the lenses in Figure 4.3.

A guided wave optical interconnect system consists of the same laser diodes and photodetectors used in the free space approach and single-level or dual-level channel waveguides. A dual-level channel waveguide system is illustrated in Figure 4.4.

Both holograms (Feldman & Guest, 1987; Turunen, Fagerholm, Vasara & Tahizadeh, 1990) and waveguides (Hickernell, 1988; Yamada, Yamada, Terui & Kobayashi, 1989) can be constructed using existing microelectronics facilities and processing methods.

Throughout this chapter, holographic and guided wave interconnects are compared in terms of their performance in implementing particular processor array connection networks. Three connection networks are considered: mesh, hypercube and fully connected. The properties of these connection networks are discussed in Section 4.2.

In Section 4.3, the efficiencies of the two optical interconnect implementations are compared for a single isolated net. The depen-

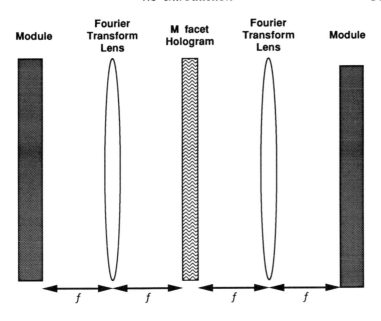

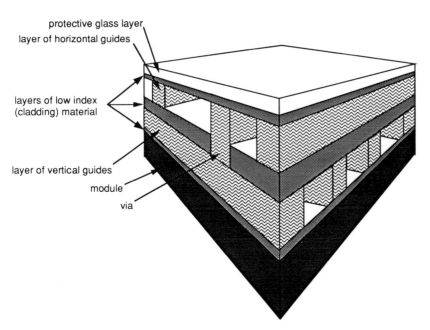

Figure 4.4 A two-level waveguide configuration as assumed in Table 4.2.

dence of the total optical link efficiency on particular variables such as fanout and waveguide attenuation is illustrated.

In Section 4.4, the connection densities of the two connection schemes are compared. Fundamental connection density limitations are compared with practical limitations for particular waveguide and hologram properties and particular processor array connection networks. It is shown that single-level guided wave interconnects are impractical for large processor arrays with high connectivity due to crosstalk limitations. In the remainder of the chapter dual-level waveguides are considered. Expressions are derived for the dependence of module area on number of processing elements (PEs).

In Section 4.5, the time delays for the two optical systems are compared. The expressions for area growth from the previous section are used to determine connection lengths and interconnect latency for particular connection networks. The power dissipations of the three connection networks for the two optical systems are compared in Section 4.6.

4.2 Connection network properties

A connection network describes the pattern of connections used to interconnect PEs within a processor array. We will assume that N PEs are evenly spaced over a square module in a rectangular array with dimensions $N^{1/2} \times N^{1/2}$. We will use the following five parameters to characterize the processor array connection network: (1) the normalized longest communication link, $L'_{\max}$; (2) the network bisection width, B; (3) the fanout from each processor, F; (4) the number of different connection patterns to be implemented, M; and (5) the minimum number of crossovers required in a single-level guided wave implementation.

The normalized longest communication link, $L'_{\max}$, is defined as

$$L'_{\max} = L_{\max} N^{1/2}/A^{1/2}, \qquad (4.1)$$

where A is the module area and $L_{\max}$ is the maximum distance separating any two directly connected PEs. If the longest connection link lies along a straight line parallel to a module edge, then $L'_{\max}$ is equal to the maximum number of PEs separating any two directly connected PEs.

The bisection width, B, is defined in terms of a partition that lies parallel to an edge of the module and divides the processor array such that $N/2$ nodes lie on either side of the partition. The bisection width is defined as the number of links crossing such a partition, minimized over all possible layouts. Note that this is a slightly different definition than that given in Thompson (1979). While Thompson's definition was designed to yield lower bounds on the layout area in terms of an asymptotic dependence on N, our definition can be used to obtain estimates on the actual layout area.

Values of normalized longest communication links (L'_{max}), network bisection widths, fanout, number of crossovers and number of different connection patterns are given in Table 4.1 for the three connection networks to be considered. Note that the larger the value for each of these parameters, the more highly connected the processor array. Thus the networks are listed in order of increasing connectivity in Table 4.1. These three networks were chosen because of their widely varying connectivity.

4.3 Optical link efficiency

The optical system efficiency, η, of an optical interconnect is defined as the ratio of the power incident on the detector (P_D) to the power emitted by the laser (P_L),

$$\eta = P_D/P_L. \tag{4.2}$$

In this section the optical link efficiencies of guided wave and free space interconnects are compared. The focus here is upon single

Table 4.1 *Parameters for three connection networks which cover a wide range of connectivity.*

Parameter	Connection network		
	Mesh	Hypercube	Fully connected
Bisection width	$2N^{1/2}$	N	$N^{3/2}$
fanout	4	$\log_2 N$	$N-1$
Crossovers	0	$\dfrac{N}{2}$	$2N^{3/2}$
L_{Max}	1	$\dfrac{N}{2^{1/2}}$	$2N^{1/2}$
Number of connection patterns (M)	4	$2\log_2 N$	$4N$

isolated nets. A net refers to a single transmitter connected to one or more receivers.

4.3.1 Free space optical interconnects

The diffraction efficiency of a hologram, η_h, is defined as the ratio of the power diffracted by the hologram to the desired location divided by the power incident on the hologram. For simplicity we will assume that the diffraction efficiencies of each subhologram in Figures 4.1 and 4.2 are identical.

For a free space optical interconnect system, if the $F\#$ of the laser subholograms matches the divergence angles of the lasers, essentially all of the light emitted by a laser illuminates the appropriate source subhologram. In this case, the optical system efficiency of a space-variant double pass system (η_{sv}) is given by

$$\eta_{sv} = \eta_h^2/F, \qquad (4.3)$$

where F is the fanout. The optical link system efficiency (η_{bs}) of a folded basis set system is

$$\eta_{bs} = \eta_h^3/F. \qquad (4.4)$$

Note that we have assumed that the diffraction efficiency of each hologram is independent of the fanout, F. This has been shown to be approximately correct by several research groups. For example in Stack & Feldman (1992), the diffraction efficiency of holograms implementing large fanout functions remains above 75% for fanout ranging from 16 to 1024.

4.3.2 Guided wave optical interconnects

The optical link efficiency for guided wave interconnects depends upon the length-dependent attenuation, $\Gamma(d)$, the source-to-waveguide coupling, κ_{sw}, and the waveguide-to-detector coupling, κ_{wd}. For unity fanout, the efficiency of a waveguide of length d is given by

$$\eta_{gw} = \kappa_{wd}\kappa_{sw}\Gamma(d), \qquad (4.5)$$

where

$$\Gamma(d) = \log^{-1}(-\gamma_d d/10), \qquad (4.6)$$

and γ_d is the waveguide loss in dB per unit length.

Two types of nonunity fanout will be evaluated. Linear fanout occurs when all destinations are distributed along a line. Linear fanout can be implemented with a single waveguide. Remote fanout, which occurs when the destinations lie in widely separated locations, requires a separate guide for each destination.

Fanout for linearly distributed destinations in a guided wave system is accomplished by power splitting a single channel. For F linearly distributed destinations addressed by a series of $F-1$ power splitters each of loss γ_s, designed so that each detector receives the same amount of incident power (neglecting length dependent attenuation effects), the power lost, P_L, due to the splitters is given by

$$P_L = 1 - \left\{ F\Gamma_s^{F-1} \Big/ \left[\sum_{k=0}^{F-2} (\Gamma_s)^k + 1 \right] \right\}, \qquad (4.7a)$$

where

$$\Gamma_s = \log^{-1}(-\gamma_s/10). \qquad (4.7b)$$

This yields a total link efficiency for a guided wave system with linear fanout of

$$\eta_{gw} = \kappa_{wd}\kappa_{sw} \left[\Gamma(d)\Gamma_s \right]^{F-1} \Big/ \left[\sum_{k=0}^{F-2} (\Gamma(d)\Gamma_s)^k + 1 \right] \qquad (4.8)$$

where d is the distance between destinations.

For remote fanout a separate waveguide is required for each location. Thus, either F sources are required, in which case there is no splitting loss, or one source can be coupled to F waveguides with the resulting splitting loss. For remote fanout with F sources the link efficiency for each destination is given by

$$\eta_{gw} = \kappa_{wd}\kappa_{sw}\Gamma(d)/F, \qquad (4.9)$$

where d is the length of travel along the waveguide from source to destination. For a remote fanout with one source a tree configuration with $\log_2 F$ couplers or a $1 \times F$ star coupler would be required.

In many cases a combination of remote and linear fanout is needed. In this case the remote fanout power loss would be calculated from Eq. (4.9) and the linear fanout power loss from Eqs. (4.7).

4.3.3 Comparison

Figure 4.5 shows the dependence of the optical link efficiency on the source-to-detector distance as calculated from Eqs. (4.3)–(4.6). The source/waveguide and waveguide/detector coupling efficiencies are assumed to be 90% each. A hologram diffraction efficiency of 85%

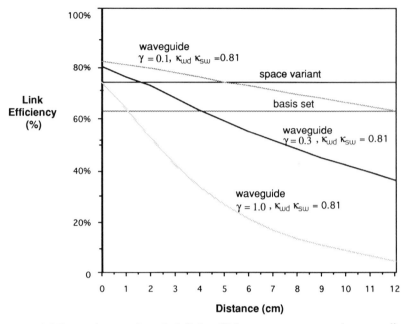

Figure 4.5 Dependence of optical link efficiency on source-to-detector distance. Coupling efficiency is neglected.

was assumed. Holographic efficiency depends upon the design, optimization method, minimum feature size, deflection angle and number of phase levels implemented (Welch & Feldman, 1990). Several research groups have fabricated holographic elements with diffraction efficiencies between 70% and 90% (Jahns & Walker, 1990; Turunen *et al.*, 1990; Welch & Feldman, 1990). Linear attenuation values ranging from 1.0 dB/cm to 0.1 dB/cm are considered in this figure. This reflects currently published values (Guha *et al.*, 1990; Hickernell, 1988; Selvaraj, Lin & McDonald, 1988). Generally, low attenuation values require thicker and wider channel waveguide structures. Attenuation results from various sources including scattering, absorption and radiation losses (Somileno, Crosignani & Porto, 1986), leakage into the substrate (Stutius & Streiter, 1977) and coupling between guides. Although noise and linear attenuation generally increase with connection density (Selvaraj *et al.*, 1988) (from increased fanout and decreased dimension), they are held constant in these comparisons. Note that for long connection lengths coupling loss represents only a small fraction of the power loss, with the remaining loss due to waveguide attenuation.

It is evident from Figure 4.5, that as the source-to-detector distance increases, the link efficiency of an optical waveguide system decreases, while the link efficiency of a free space system remains constant. This is analogous to comparisons between free space and electrical inter-connect systems (Feldman, Esener, Guest & Lee, 1988), where for lengths longer than a particular break-even line length it is advantag-eous to use optical interconnects. Figure 4.5 shows that similarly, when comparing free space and waveguide interconnects, there is a break-even distance for which free space interconnects have superior per-formance. This break-even distance will depend on the hologram efficiency, waveguide attenuation and coupling losses, but for current state-of-the-art waveguides and holograms, this distance is approxi-mately 5 cm.

Figure 4.6 shows the effect of linear fanout on interconnect efficien-cies using Eqs. (4.3) and (4.8). This example assumes constant splitter loss for each guide. Low splitter loss values of 1.2 dB and 1.8 dB (Chung, Spickermann, Young & Dagli, 1992) are used in the figure. Attenuations of 0.1 dB/cm (Selvaraj *et al.*, 1988), 0.3 dB/cm (Guha *et*

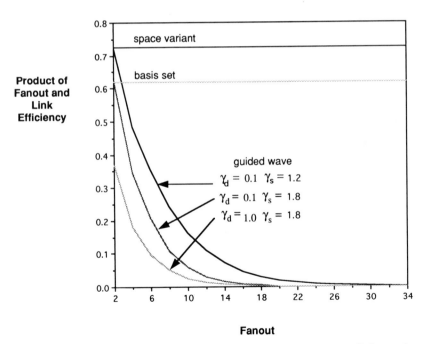

Figure 4.6 Dependence of optical link efficiency for the longest link on the number of linear fanout destinations. The connection length is 5 cm.

al., 1990) and 1.0 dB/cm (Hickernell, 1988) are shown. The connection length is 5 cm.

4.4 Connection density

4.4.1 Fundamental connection density limitations

In the following section the fundamental limits on connection density capacities are calculated. The maximum number of waveguides which can cross a given width $(w = A)^{1/2}$ is determined. This is compared to the number of connections which can be made with a hologram of area A.

Guided wave optical interconnects

Guided wave systems are limited by size and spacing considerations (Haugen, Rychnovsky, Husain & Hutcheson, 1986; Kogelnik, 1981). The minimum dimension for confinement to occur depends upon the design wavelength and the difference between the refractive indices of the material of the guide, n_w, and of the cladding surrounding the waveguide, n_c. For confinement to occur the waveguide core diameter, W, must have a minimum dimension of (Haugen *et al.*, 1986),

$$W = \lambda/(n_w^2 - n_c^2)^{1/2}. \tag{4.10}$$

The spacing between guides is limited by crosstalk considerations. The ratio of the amount of power coupled to a neighboring guide to the power remaining in an excited waveguide is referred to as the power ratio. The power ratio between two identical waveguides that are parallel for a distance L is

$$PR = L^2 C^2, \tag{4.11}$$

where C is the coupling coefficient (Marcuse, 1982b). The coupling coefficient between two identical waveguides of width W and index n_g surrounded by material of index n_c and separated by a distance S_g is (Marcuse, 1982a)

$$C = \frac{(n_g^2 k_0^2 - \beta^2)(\beta^2 - n_c^2 k_0^2)}{\beta(1 + \gamma d)k_0^2(n_g^2 - n_c^2)} e^{\gamma(W - s_g)} \tag{4.12}$$

where

$$k_0 = 2\pi/\lambda_0, \ \gamma = (k_0^2 n_c^2 - \beta^2)^{1/2},$$

and

$$\beta^2 = n_{\rm g}^2 k_0^2 \left[1 - \frac{2(\Delta n)^{1/2}}{n_{\rm g} k_0 W} \right],$$

where

$$\Delta n = (n_{\rm g}^2 - n_{\rm c}^2) n_{\rm g}^2.$$

The total number of connections is represented by $\#c$. The number of connections is equal to the product of the number of elements to be connected, N, and their respective fanout, F. The total connection density in number of connections per unit length is

$$\#c/A = 1/(S + W) = [S_{\rm g} + \lambda/(n_{\rm w}^2 - n_{\rm c}^2)^{1/2}]^{-1}. \tag{4.13}$$

Thus the maximum bisection width that can be implemented is given by

$$B \leq A^{1/2}/(S_{\rm g} + W). \tag{4.14}$$

The number of processors that can be interconnected can then be determined from the bisection width of the connection network.

Free space optical interconnects

The fundamental limits on the density capabilities of holographic interconnects can be determined from diffraction-limited beam divergence properties. Maximum connection density can be achieved with a space-invariant optical system. In a space-invariant system every PE employs the same connection pattern. This can be achieved by choosing a connection pattern that is a superset of all desired connection patterns and masking off undesired connection links. This can be implemented with a basis set optical system (Figures 4.2 and 4.3) with the number of basis set connections, M, set equal to 1.

The area of the hologram $(A_{\rm h})$ may be found by totaling the area of each source subhologram $(A_{\rm s})$ and detector subholograms $(A_{\rm d})$. Setting the two areas equal optimizes total area. This yields

$$A_{\rm h} = N A_{\rm s} + NF A_{\rm d} = 2NF A_{\rm d} \tag{4.15}$$

for the total hologram area. The area of each detector subhologram is limited by (Jenkins *et al.*, 1984)

$$A_{\rm d} \geq 4 (1.2)^2 \lambda^2 f^2/A \cos^4 \Phi \tag{4.16}$$

where f is defined in Figure 4.2 and Φ is the deflection angle.

Since the PEs are uniformly distributed, the subhologram area associated with a given PE is identical for every PE in the network. To determine the lower bound on connection density set $A_{\rm d}$ to the minimum area of the largest subhologram. The size of each detector subhologram is given by the value of A_d for a subhologram at the edge of the network for which $\tan \Phi = A^{1/2}/2f$, yielding

$$A_d = (1.2)^2 \lambda^2 / \cos^4 \Phi \tan^2 \Phi. \qquad (4.17)$$

A_d is minimized for $\Phi \approx 45°$, resulting in a lower bound on the hologram area of

$$A \geqslant 4 \, (1.2)^2 \lambda^2 NF. \qquad (4.18)$$

A maximum connection density is achieved for a bisection width, B, equal to NF, for which the number of connections per unit width is given by

$$B/A^{1/2} = A^{1/2}/4(1.2)^2 \lambda^2. \qquad (4.19)$$

Comparison

While the connections per unit width increases with area for free space interconnects, it is independent of area for guided wave systems. The maximum number of connections per millimeter of module width can be calculated using Eqs. (4.13) and (4.19) for guided wave and free space interconnects, respectively. The waveguide width can be set to minimum width for containment according to Eq. (4.10). For both free space and guided wave interconnects, the connection density increases with increasing index of refraction. 'Index of refraction' refers to the core index for the waveguide and the index between the module surface and the reflecting hologram in Figure 4.2 for the free space case. However, increasing index of refraction values results in a slower optical propagation speed for both guided wave and free space systems. For this reason, both optical technologies tend to employ low index of refraction materials (typically $1.0 < n < 2.0$). In addition, the use of high index of refraction materials in the free space case can result in additional losses due to Fresnel reflections. These can be minimized through the use of anti-reflection coatings (see Welch, Morris & Feldman (1993) where a silicon hologram with index of refraction of 3.5 was fabricated with a net Fresnel loss on both surfaces of less than 3%) and planarizing techniques.

4.4.2 Practical limitations

Guided wave optical interconnects

The large number of crossovers in a single-level guided wave system used to implement a highly connected network can drastically reduce the connection density. The number of crossovers as a function of number of processors for several interconnect configurations may be found in Table 4.1. Consider the system efficiency of a single-level guided wave system. If γ_d is the linear attenuation and γ_x is the crossover loss (in dB per unit length), then the link efficiency is

$$\eta_{gw} = \kappa \log^{-1}(-\gamma_d d/10) \log^{-1}(-\gamma_x x/10). \qquad (4.20)$$

Next consider the signal to noise ratio (SNR) for a single-level guided wave interconnect system. In this analysis we will consider only the signal and noise inherent to waveguide systems that are not present for free space systems. This is the reduced signal due to attenuation and increased noise due to power leakage between waveguides from crossovers. (Of course the actual SNR will be smaller than these values due to additional noise sources in common to both free space and waveguide systems, such as noise resulting from optical quantum effects, and noise associated with the transmit and receive electronics.) Consider S the signal power remaining in a waveguide after the signal has traveled a distance d and crossed x other guides. Let γ_d be the power lost per crossover. Let $noise_x$ be the power that is coupled to the waveguide under consideration from the guide perpendicular at each crossover. Assume identical guides. N is the sum of power which has been transferred into the waveguide during crossovers. The ratio of power remaining (S) to noise from crossovers (N) follows:

$$S = \log^{-1}(\gamma_x x) \log^{-1}(\gamma_d d), \qquad (4.21)$$

$$N = x \log^{-1}(noise_x), \qquad (4.22)$$

and

$$\mathrm{SNR} = \frac{\log^{-1}(\gamma_x x) \log^{-1}(\gamma_d d)}{x \log^{-1}(noise_x)}. \qquad (4.23)$$

Figure 4.7 shows the number of processors which could be connected in a hypercube according to basic efficiency and noise limits for single-level guided wave as described by Eqs. (4.14), (4.20) and (4.23). The physical limit for guides of index 2.0 is nearly 10^5 possible connections for a 10 cm module diameter if 10% coupling loss between guides can be tolerated. Values of 0.1 dB/cm attenuation and 0.006 dB/crossover loss were used for the efficiency and SNR limits in Figure 4.7. If a minimum requirement of 10% system efficiency is imposed, only a fraction ($N < 4000$) of the basic limit would be reached. Note that the electrical-to-optical and optical-to-electrical

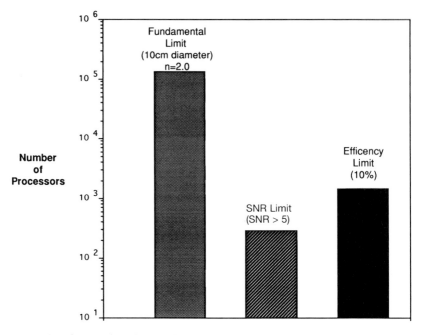

Figure 4.7 Connection density limitations for waveguide optical interconnects. The fundamental limit on the number of connections in a 10 cm module diameter for guide index 2.0, cladding index 1.45, and 10% crosstalk loss is shown using Figure 4.5. The number of connections which can be made in a hypercube if the tolerable loss is 10%, and the number of connections which can be connected if the lowest tolerable SNR is 5 are shown.

conversion efficiencies are not included in our definition of optical system efficiency. This is because these efficiencies are common to free space and guided wave systems. Given a system which has -52 dB/crossover crosstalk and can tolerate an SNR of not less than 5, only a few processors can be connected.

Figure 4.8 shows the number of processor elements that could be connected as a function of tolerable crosstalk as determined from Eq. (4.12) and Table 4.1 for fully connected and hypercube networks. The same values for attenuation and crosstalk per crossover were used as in Figure 4.7. Note that only a few PEs can be reasonably connected. The severe limitations on connection density illustrated in Figure 4.7 and 4.8 are primarily due to the large number of guided wave crossovers encountered in single-level waveguide interconnect systems. If a mesh or any other crossover-free network is implemented, the connection density is limited only by Eqs. (4.10)–(4.14).

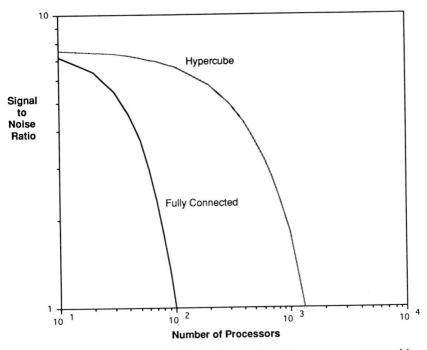

Figure 4.8 The dependence of SNR on the number of processors connected in a hypercube and fully connected networks. The SNR corresponds to that of the longest connection in an array of a given number of processors.

If single-level guided wave interconnects are to be used in highly connected, large size processor arrays, a multiple layer implementation is necessary. Therefore, for the remainder of the chapter all comparisons between guided wave and free space systems will be based on a dual-level guided wave system where appropriate. A two-layer system consisting of layers of waveguides where each guide is parallel to all guides on the same level and perpendicular to all guides on a different level is shown in Figure 4.4. By routing signals between the two levels, any connection network can theoretically be implemented without any crossovers. However, the implementation of such dual-level guided wave systems, including coupling signals between guides in different levels is a difficult technological problem that to date has not been solved.

Since theoretically dual-level guided wave systems contain no crossovers, the interconnect density of such systems is limited by Eq. (4.14). Eq. (4.12) defines the minimum waveguide pitch which allows tolerable crosstalk and loss. However, note that Eqs. (4.12) and (4.14) are based on ideal waveguides and suggest the capability to construct arbitrarily small waveguides that are virtually defect-free. This capability does not currently exist. As the waveguide diameter is made smaller, the effects of defects become catastrophic. A small defect can cause a nearly total loss of confinement. As the length of the waveguide increases, the probability of a waveguide containing such a defect increases.

Comparisons of systems will consider waveguide pitches from published results. The best currently published results include ridge waveguides with guide widths of 4 μm (Baba & Kokobun, 1990), 5 μm (Eck *et al.*, 1994), and 9 μm, (Mak, Bruce & Jessop, 1989) which would yield pitches of approximately 8 μm, 10 μm, and 18 μm respectively. Since connection lengths of ~10 cm are needed for chip-to-chip interconnects, in the remaining graphical examples values of waveguide pitch ranging from 4 μm to 10 μm are used to illustrate practical limitations. The minimum pitch case is also presented.

Free space optical interconnects

The fundamental connection density limit given in Section 4.4.1 was derived for a space-invariant connection system in which the light from each laser beam illuminated an aperture with an area equal to the area of the multi-chip module. Such optical systems allow for the maximum connection density. However, for many connection networks, optical systems that provide such a connection density have not been found. Thus, in this section the connection densities of the double pass space-variant and folded basis set systems are determined.

Space-variant systems   The double pass space-variant optical interconnect system is illustrated in Figure 4.1. The total hologram area is related to the detector subhologram area by Eq. (4.15). The detector subhologram area is given by (Feldman & Guest, 1989)

$$A_d = 16 \, (1.2)^2 \lambda^2 h^2 / \{D^2 \cos^4 \Phi\}, \tag{4.24}$$

where D is the side length of the source subhologram. Using Eq. (4.15),

$$D^2 = FA_d. \tag{4.25}$$

Substituting Eq. (4.25) into Eq. (4.24) and noting that tan $\Phi = L_{max}$ 2h yields

$$A_d = 2 \, (1.2)\lambda L_{max}/F^{1/2} \cos^2 \Phi \tan \Phi, \tag{4.26}$$

where L_{max} is the length of the longest connection. Using Eq. (4.15) and Eq. (4.1) yields

$$A = 16 (1.2)^2 \lambda^2 L_{max}'^2 \, NF/\cos^4 \Phi \tan^2 \Phi. \qquad (4.27)$$

Again using $\Phi = 45°$ yields

$$A \approx 92 NF \lambda^2 L_{max}'^2. \qquad (4.28)$$

For a mesh, $L_{max}' = 1$, yielding a connection density, $B/A^{1/2}$, of $\sim 1/19\lambda$. For a hypercube, $L_{max}' = N^{1/2}/2$, yielding a connection density, $B/A^{1/2}$ of $\sim (4.8 F^{1/2} \lambda)^{-1}$.

Basis set system A folded basis set hologram configuration is shown in Figure 4.3. The area for the hologram in a folded basis set can be calculated in a manner similar to the above derivation. This results in the following expression:

$$A \approx 92 NFM\lambda^2, \qquad (4.29)$$

where M is the number of different connections in the connection network or the number of basis set subholograms.

Space-invariant system For a fully connected network, and all networks with very high connectivity, a basis set offers no advantage because of the number of connections needed. A space-invariant system, where each source is connected to a $2N \times 2N$ array can be used to reduce the hologram area greatly. However, the optical system efficiency of such a system is reduced by a factor of 4. The area of a space-invariant system can be calculated by setting M equal to 1 in Eq. (4.29).

Comparison

Figures 4.9(a)–(c) show the total processor area, the total area of waveguides and the total hologram area for the three interconnect networks. Figure 4.9(a) shows the connection network area and processor areas for a processor array connected in a mesh. Figure 4.9(b) shows the same quantities for an array connected in a hypercube. Similarly, Figure 4.9(c) shows the connection network and processor areas for an identical processor array that is fully connected. The array is of $N \times N$ processors each of width equal to 0.35 mm. The total area for the waveguides is calculated from Eq. (4.14). Eqs. (4.26) and (4.27) were used to determine the CGH area.

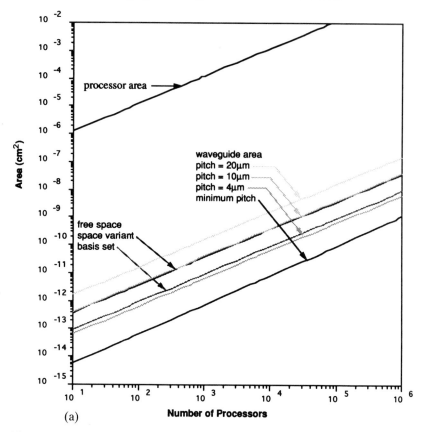

Figure 4.9 Processor area, area required by waveguides, and hologram area for connecting processors in (a) a mesh interconnect network, (b) a hypercube configuration, and (c) a fully connected system.

The effects of differing guide pitches and holographic implementations are clear in Figures 4.9(a)–(c). These figures show that for a particular number of processors the total area of the waveguides can become greater than the total area of the processing elements. This occurs between 10 and 200 processors (depending on waveguide width) in a fully connected network, between 100 and ~50 000 processors in a hypercube network and not at all in a mesh interconnect network. Note that space-variant holographic systems have a connection capacity similar to that of a two-level waveguide system with a 10 μm pitch for all three connection networks. For basis set and space-invariant systems the area of the CGH never exceeds the processor area for all cases considered.

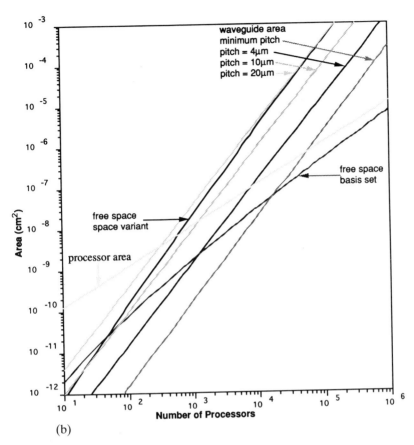

(b)

In Figure 4.9, destinations were assumed to be remote so that there is at least one waveguide per destination. It was also assumed that data are transmitted in serial, so there is at most one waveguide per destination.

4.5 Interconnect delay time

In this section expressions for the time delay for guided wave and free space connections for identical source-to-detector distances are derived.

The time delay in a free space interconnect is proportional to the distance the signal travels, d; R is the mirror-to-hologram distance and

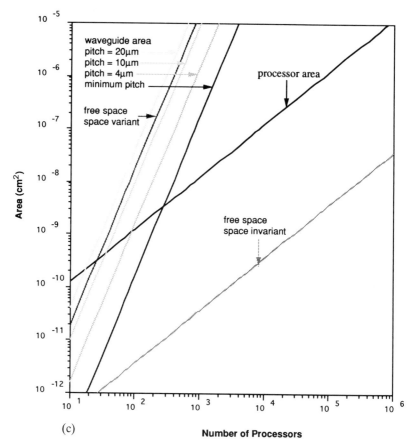

(c)

Number of Processors

Figure 4.9 *Continued*.

L is the source-to-detector distance. As a function of R, d is

$$d = 2[R^2 + (d/2)^2]^{1/2}. \tag{4.30}$$

The hologram–mirror separation required to make a connection of distance L is $R \geqslant L/2$. Substituting minimum R yields a time delay of

$$\tau_{fs} = 2^{1/2} L/c. \tag{4.31}$$

The velocity, v, within a waveguide is determined by the intrinsic impedance, n, of the waveguide material. The wave velocity is limited by c/n.

The source-to-detector distance, L, is related to waveguide length d by a scaling factor α. This scaling factor accounts for the increased distance of connections in a system with planar constraints. The time delay τ_{gw} is given by

$$\tau_{gw} = \alpha Ln/c. \tag{4.32}$$

The scaling factor, α, can vary between 1 and $2^{1/2}$ for a dual-level waveguide system.

4.5.1 Comparisons

The longest connection length for the three connection networks can be found in Table 4.2. Except for the mesh connection network, each enty in Table 4.2 has two expressions. The larger of the two expressions is the effective value. If the area of the PEs is larger than the area of the interconnect medium, then the first expression is valid. When the area of the interconnection medium is larger than the area of the PEs, the second expression will be larger and thus the determinant of the connection length. The crossover points are illustrated in Figures 4.9(a)–(c) for the three networks.

Examination of Eqs. (4.30) and (4.32) shows that the ratio of τ_{gw} to τ_{fs} is $\alpha n L_{gw}/2^{1/2} L_{fs}$, where L_{gw} is the source-to-detector separation for guided wave interconnects and L_{fs} is the source-to-detector separation for free space interconnects. The distance scaling term, α, depends on the connection network and optical system employed. For example, the ratio of guided wave optical path to basis set free space optical path for a large number of processors in a hypercube is given by

Table 4.2 *Implementation-dependent connection configuration parameters for two-level waveguide and double pass hologram implementations.*

Connection network	Implementation	Longest connection length	Interconnect medium width
Mesh	Guided wave	W_{pe}	$2 N^{1/2} W_{wg}$
	Free space	W_{pe}	$A_h^{1/2}$
Hypercube	Guided wave	$\dfrac{N^{1/2} W_{pe}}{2}$ or $\dfrac{N W_{wg}}{2}$	$N W_{wg}$
	Free space	$\dfrac{N^{1/2} W_{pe}}{2}$ or $\dfrac{A_h}{2}$	$A_h^{1/2}$
Fully connected	Guided wave	$2 N^{1/2} W_{pe}$ or $2 N^{3/2} W_{wg}$	$N^{3/2} W_{wg}$
	Free space	$(2N)^{1/2} W_{pe}$ or $(2A_h)^{1/2}$	$A_h^{1/2}$

$$N^{1/2} \frac{W_{wg}}{W_{pe}} \frac{n}{2^{1/2}}. \qquad (4.33)$$

Eq. (4.33) was determined from Table 4.2 and Eq. (4.32). Since it is evident from Figures 4.9(a)–(c) that the basis set hologram area never exceeds the PE area for $N < 10^6$, the first expression in Table 4.2 for the free space longest connection length was used. On the other hand, for $N > \sim 5000$ the waveguide area exceeds the PE area, which is why the second expression in Table 4.2 for the waveguide longest connection length was used.

Figures 4.10(a) and (b) illustrate the time delay for the longest length connection in the processor array for a hypercube and fully

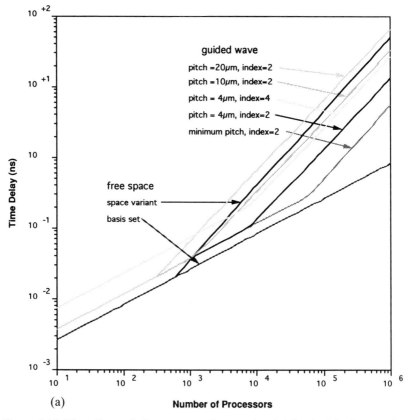

(a)   **Number of Processors**

Figure 4.10 The effects of planar constraints on signal delay in (a) a hypercube and (b) a fully connected network.

connected configuration. Eqs. (4.31) and (4.32) and Table 4.2 are used.

The change in the rate of increase of time delay occurs at the point at which the area of the interconnects becomes greater than the module area. Note that the processors and the interconnects are on different planes so that this occurs at a discrete point. After this point the rate of increase of the time delay depends upon the rate of increase of the connection network area. Before this point the slope of the time delay curve is equal to the rate of increase of the processor area.

As the number of processing elements reaches the thousands, the ratio of optical paths produces a dramatic differential in speed capacities between free space basis set and guided wave optical implemen-

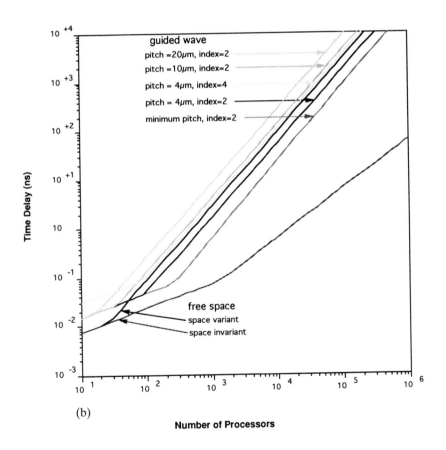

(b)

Number of Processors

tations. This difference appears for a lower number of processors and becomes greater more quickly for waveguides with larger pitches. No' that the free space space-variant system has a time delay similar to th of a waveguide system with a refractive index of 4.0 and a pitch c 4 μm.

4.6 Power dissipation

The switching energy for an arbitrary optical interconnect system is given by (Feldman *et al.*, 1988)

$$E = 2TP_{th} + 2(C_{gate} + C_{pd})V_{cc}\frac{h\nu}{\eta q}. \tag{4.34}$$

Eq. (4.34) is the sum of the laser threshold power, P_{th}, and the necessary optical power to drive the detector. C_{gate} and C_{pd} are the capacitances of the gate and the photodetector respectively. η represents the total optical link efficiency, h represents Planck's constant, ν is the optical frequency, q is the electronic charge and V_{cc} is the power supply voltage. The factor of 2 in both components is to account for an entire high-to-low cycle. A duty cycle of one is assumed. Both the charging and discharging of the receiving gate are considered.

For waveguide interconnects this expression becomes

$$E_{gw} = 2P_{th}T + 2(C_{gate} + C_{pd})\frac{h\nu}{q}\frac{V_{cc}}{\kappa_{wd}\kappa_{sw}\eta_s\eta_d a \log(\gamma \cdot 1)}. \tag{4.35}$$

A similar expression for free space interconnects is given by

$$E_{fs} = 2P_{th}T + 2(C_{gate} + C_{pd})\frac{h\nu}{q}\frac{V_{cc}}{\eta_s\eta_d\eta_h}. \tag{4.36}$$

Figure 4.11(a) show the switching energy for a connection in a mesh connection network. Figures 4.11(a)–(c) shows the switching energy for the longest length connection for mesh, hypercube and fully connected networks, respectively. These plots were obtained from Table 4.2 and Eqs. (4.35) and (4.36). Eqs. (4.3) and (4.4) were used for calculating the optical link efficiency of the free space systems. Eqs. (4.5) and (4.6) were used in calculating the optical system efficiency for the guided wave interconnect systems. The length of the longest connection can be determined from Table 4.2. Only the longest link is considered, so the fanout is neglected. A core refractive index value of 2.0 was used for these plots.

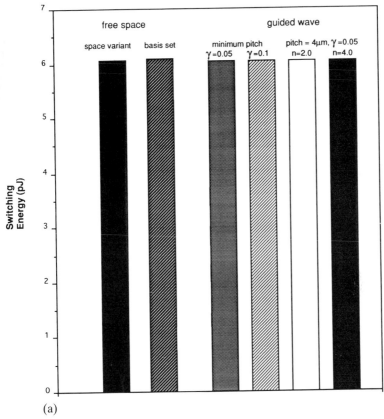

(a)

Figure 4.11 Switching energy for: (a) a connection in a mesh configuration; (b) the longest connection for up to 100 K processors connected in a hypercube configuration; and (c) the longest connection for an array of up to 100 K processors in a fully connected network. For all graphs $\kappa_{wd}\kappa_{sw} = 85\%$, $\eta_d\eta_s = 0.4$ and $\eta_h = 85\%$, unless otherwise stated.

Values of 1 GHz for the data rate, 3.0 mW for the laser threshold power, 2.0 V for the electrical power supply, and a total capacitance of 5.97 fF were employed. These values are typical for a 0.8 μm complementary metal oxide semiconductor (CMOS) process. As in the previous numerical examples a hologram efficiency of 85% is assumed.

Note that the switching energy for free space systems is independent of the number of PEs. The switching energy for guided wave systems increases with increasing numbers of processors due to their length-dependent attenuation. While for a small number of processors in networks with a low level of connectivity guided wave systems may have a lower switching energy, as the number of processors increases

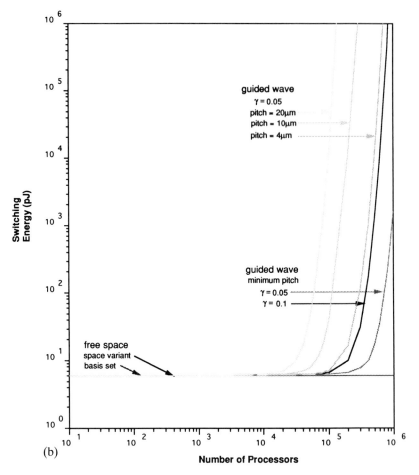

Figure 4.11 *Continued.*

and/or the connectivity level increases, the switching energy of free space interconnects becomes significantly lower than that of guided wave systems.

4.7 Conclusions

In this chapter holographic interconnects were compared to free space interconnects in terms of connection density, interconnect latency and power dissipation.

Results indicate that single-level guided wave interconnects are impractical for highly connected large scale processor arrays due to

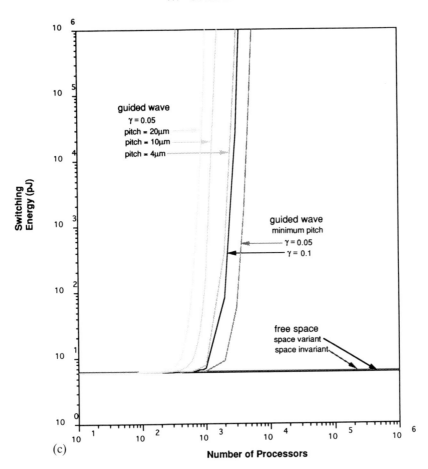

(c)

crosstalk and power loss occurring at guide crossover points. Although high performance dual-level wave guide systems have not been demonstrated, they have the potential to increase performance greatly since all guide crossover points may be eliminated in such systems. Figures 4.9–4.11 compare the performance of holographic interconnect systems to dual-level interconnect systems.

The time delay and connection density of a holographic space-variant system are comparable to those of a two-level guided wave system with a pitch of ~10 μm (width of ~5λ where λ is 1 μm) and core refractive index of approximately 2.0 (Figures 4.9 and 4.10). The power dissipation of guided wave systems can be kept comparable to free space interconnect systems if the waveguide attenuation can be kept below ~0.5 dB/cm and the number of PEs is kept below ~100 000.

For highly connected large scale processor arrays, free space basis set and space-invariant optical systems can attain a performance advantage of several orders of magnitude over both guided wave and free space space-variant systems in terms of connection density, time delay and power dissipation (Figures 4.9–4.11). This performance advantage increases with increasing network connectivity (as described by the parameters in Table 4.1) and as the number of PEs in a network increases. For $N > {\sim}10\,000$ (the actual number will depend on the specific PE dimensions), such optical systems can reduce the interconnect delay of the longest link in a connection network by at least one order of magnitude below fundamental limits on dual-level guided wave optical interconnect systems.

In discussing packing density in this chapter, we have ignored several potential limiting factors including noise due to crosstalk between closely spaced lasers, and for free space interconnects crosstalk due to power in undesired diffraction orders. While care must be taken to separate laser sources by sufficient distances to avoid crosstalk, this can be done in a manner that would not increase the system area significantly for virtually all of the cases presented in this chapter. For example, for closely spaced, guided wave interconnects, lasers could be distributed along the length of the waveguides. For free space interconnects, as the number of processors in the system grows, the center-to-center spacing between holograms also grows, resulting in increasing separations between lasers and therefore decreasing laser-to-laser proximity crosstalk. Similarly, it was shown in Schwider, Stork, Streibl & Völkel (1992), that crosstalk in holographic interconnect systems resulting from power contained in undesired diffraction orders decreases with increasing system size. However, this analysis was based on the use of a particular type of hologram and the assumption that the only significant noise power was contained in the unfocused zeroth diffraction order of each hologram facet. When large fanouts are implemented it is possible to couple power into highly focused diffraction orders that could significantly reduce the SNR. Nevertheless, it is also possible to design a system so that these highly focused spots do not coincide with any detectors.

An additional issue not raised in this chapter is the yield and reliability of these systems. It is our belief that holographic interconnects have a significant advantage in this area over channel waveguide systems. For example, from our analysis, a space-variant free space system had an area comparable to that of a guided wave system with channel width of 5λ. Yet as the area of the system grows, the probability that a defect in the fabrication process of the waveguide will cause a critical failure will increase, since the channel width is kept constant while the length increases. On the other hand, with the space-variant holographic system, the size of the hologram facets increases with increasing area thereby decreasing the failure probability due to defects in the holograms as the system grows.

References

Baba, T., and Kokobun, Y., High Efficiency Light Coupling from Antiresonant Reflecting Optical Waveguide to Integrated Photodetector using an Antireflecting Layer. *Applied Optics*, **29**, 2781–91 (1990).

Chung, Y., Spickermann, R., Young, D. B., and Dagli, N., A Low-Loss Beam Splitter with an Optimized Waveguide Structure. *IEEE Photonics Tech. Lett.*, **4**, 1009–11 (1992).

Eck, T. E. V., Ticknor, A. J., Lytel, R., and Lipscomb, G. F., A Complementary Optical Tap Fabricated in an Electro-Optic Polymer Waveguide. to be published in *Appl. Phys. Lett.*, (1994).

Feldman, M. R., Esener, S. C., Guest, C. C., and Lee, S. H., Comparison between Optical and Electrical Interconnects based on Power and Speed Considerations. *Applied Optics*, **27**, 1742–51 (1988).

Feldman, M. R., and Guest, C. C., Computer Generated Holograms for Optical Interconnection of VLSI Circuits. *Applied Optics*, **26**, 4377–84 (1987).

Feldman, M. R., and Guest, C. C., Interconnect Density Capabilities of Computer Generated Holograms for Optical Interconnection of Very Large Scale Integrated Circuits. *Applied Optics*, **28**, 3134–7 (1989).

Feldman, M. R., Guest, C. C., Drabik, T. J., and Esener, S. C., Comparison between Electrical and Free-Space Optical Interconnects for Fine Grain Processor Arrays based on Interconnect Density Capabilities. *Applied Optics*, **28**, 3820–9 (1989).

Frietman, E. E. E., Nifterick, W. V., Dekker, L., and Jongeling, T. J. M., Parallel Optical Interconnects: Implementation of Optoelectronics in Multiprocessor Architecture. *Applied Optics*, **29**, 1161–7 (1990).

Guha, A., Briston, J., Sullivan, C., and Husain, A., Optical Interconnects for Massively Parallel Architectures. *Applied Optics*, **29**, 1077–93 (1990).

Haugen, P. R., Rychnovsky, S., Husain, A., and Hutcheson, L. D., Optical Interconnects for High Speed Computing. *Optical Engineering*, **25**, 1076–85 (1986).

Hickernell, F., Optical Waveguides on Silicon. *Solid State Technology*, **31**, 83–7 (1988).

Jahns, J., and Walker, S. J., Two-Dimensional Array of Diffractive Microlenses Fabricated by Thin Film Deposition. *Applied Optics*, **29**, 931–6 (1990).

Jenkins, B. K., Chavel, P., Forchheimer, R., Sawchuck, A. A., and Strand, T. C., Architectural Implications of a Digital Optical Processor. *Applied Optics*, **23**, 3465–74 (1984).

Kogelnik, H., Limits of Integrated Optics. *Proceedings of the IEEE*, **69**, 232–8 (1981).

Mak, G., Bruce, D., and Jessop, P., Waveguide Detector Couplers for Integrated Optics and Monolithic Switching Arrays. *Applied Optics*, **28**, 4505–636 (1989).

Marcuse, D., Coupling Between Dielectric Waveguides. In *Light Transmission Optics*: Second Edition, B. T. L. Inc., Van Nostrand Rienhold Publishing, New York, NY., p. 427 (1982a).

Marcuse, D., Coupling Between Dielectric Waveguides. In *Light Transmission Optics*: Second Edition, B. T. L. Inc., Van Nostrand Reinhold Publishing, New York, NY, p. 425 (1982b).

Schwider, J., Stork, W., Streibl, N., and Völkel, R., Possibilities and Limitations of Space-Variant Holographic Elements for Switching Networks and General Interconnects. *Applied Optics*, **31**, 7403 (1992).

Selvaraj, R., Lin, H. T., and McDonald, J. F., Integrated Optical Waveguides in Polyimide for Wafer Scale Integration. *IEEE/OSA Journal of Lightwave Technology*, **6**, 1034–7 (1988).

Somileno, S., Crosignani, B., and Porto, P. D., *Guiding Diffraction and Confinement of Optical Radiation*. Academic Press, London, p. 563 (1986).

Stack, J. D., and Feldman, M. R., Recursive Mean-Squared-Error Algorithm for Iterative Discrete On-Axis Encoded Holograms. *Applied Optics*, **31**, 4839–46 (1992).

Stirk, C. W., Athale, R. A., and Haney, M. W., Folded Perfect Shuffle Optical Processor. *Applied Optics*, **27**, 202–3 (1988).

Stutius, W., and Streiter, W., Silicon Nitride Films on Silicon for Optical Waveguides. *Applied Optics*, **16**, 3218–22 (1977).

Thompson, C. D., Area-Time Complexity for VLSI. *Proceedings of the Eleventh Annual ACM Symposium on the Theory of Computing*, Association for Computing Machinery, New York, NY, pp. 81–8 (1979).

Turunen, J., Fagerholm, J., Vasara, A., and Tahizadeh, M., Detour Phase Kinoform Interconnects: The Concept and Fabrication Considerations. *Journal of the Optical Society of America*, **29**, 1202–8 (1990).

Welch, W. H., and Feldman, M. R., Iterative Discrete On-Axis (IDO) Encoding of Diffractive Optical Elements. *OSA Annual Meeting, Technical Digest Series*, **15**, 72 (1990).

Welch, W. H., Morris, J. E., and Feldman, M. R., Iterative Discrete On-axis Encoding of Radially Symmetric Computer Generated Holograms. *Journal of the Optical Society of America–A*, **10**, 1729–38 (1993).

Yamada, Y., Yamada, M., Terui, H., and Kobayashi, M., Optical Interconnections Using a Silica-Based Waveguide on a Silicon Substrate. *Optical Engineering*, **28**, 1281–7 (1989).

5

High speed compact optical correlator design and implementation

RICHARD M. TURNER and KRISTINA M.
JOHNSON
University of Colorado, Optoelectronic Computing Systems Center

and STEVE SERATI
Boulder Nonlinear Systems, Inc.

5.1 Introduction to optical correlator architectures

Optical pattern recognition has been used in such applications as optical inspection (Applied Optics Special Edition, 1988), target and character recognition (Tippett *et al.*, 1965), synthetic aperture radar processing (SAR) (Cutrona, Leith, Porcello & Vivian, 1966), and image processing. The main optical processing subsystem in these applications is the 2-D optical correlator. Recently there has been a renaissance in optical correlators due to the development of high speed spatial light modulators (SLMs). In this chapter, we review the mathematics of performing optical correlations, discuss SLM technology and the influence of SLM characteristics on the correlator system, describe the design of high speed optical correlators using SLMs, and discuss the future of SLMs and optical correlator systems.

5.1.1 Background

The optical correlation between two images, $f(x, y)$ and $s(x, y)$, is written as

$$c(x, y) = f(x, y) \star s(x, y), \qquad (5.1)$$

where the $\star$ symbol represents the correlation operation. Employing Fourier transforms (Bracewell, 1965) allows Eq. (5.1) to be rewritten as

$$c(x, y) = \mathcal{F}\mathcal{J}\{F(f_x, f_y)S^*(f_x, f_y)\}, \qquad (5.2)$$

where $F(f_x, f_y)$ is the Fourier transform ($\mathcal{F}\mathcal{J}$) of $f(x, y)$ and $S^*(f_x, f_y)$ is the complex conjugate of the Fourier transform of $s(x, y)$. There are two well-known optical architectures for finding $c(x, y)$: the joint transform correlator (JTC) and the VanderLugt correlator (Weaver & Goodman, 1966; VanderLugt, 1964). In the JTC architecture, shown in Figure 5.1, spatially separated images $f(x, y)$ and $s(x, y)$ are simultaneously displayed on an input electrically addressable SLM (EASLM). A lens takes an analog Fourier transform of the two images which is then recorded on an optically addressable SLM (OASLM). The intensity distribution of the transform is given by

$$|F(f_x, f_y) \exp(i2\pi f_y y_1) + S(f_x, f_y) \exp(i2\pi f_y y_2)|^2, \qquad (5.3)$$

where y_1 and y_2 represent the offsets of $f(x, y)$ and $s(x, y)$ from the origin in the EASLM. This recorded intensity distribution is then read off of the OASLM and optically Fourier transformed yielding three spatially separated terms in the correlation plane,

$$f(x, y) \star f(x, y) + s(x, y) \star s(x, y), \qquad (5.4)$$

$$f(x, y) \star s[x, y - (y_1 - y_2)], \qquad (5.5)$$

$$s(x, y) \star f[x, y + (y_1 - y_2)], \qquad (5.6)$$

corresponding to the autocorrelation function of $f(x, y)$ and $s(x, y)$ located at the origin of the output plane, and the cross-correlations of $f(x, y)$ and $s(x, y)$ centered at $y_1 - y_2$ and $-(y_1 - y_2)$, respectively. Provided the spacing in the correlation plane is adequate to separate the three terms, the desired cross-correlation information, $c(x, y) = f(x, y) \star s(x, y)$, is easily obtained by spatial filtering. Separating the two inputs by a distance equal to the width of the larger of the two input images will result in nonoverlapping output signals.

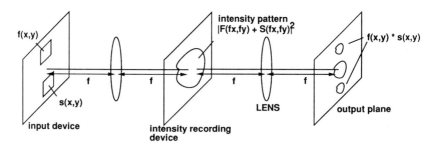

Figure 5.1 Typical JTC layout. The input device is an SLM containing both input and filter functions. The filter plane device is a photosensitive SLM which records the transform of the input SLM. Readout of the filter plane device and subsequent Fourier transform yields the correlation function.

The JTC architecture requires twice the space-bandwidth capacity of an equivalent VanderLugt correlator. It also requires a square-law device for the detection and display of the intermediate intensity pattern (see Eq. (5.3)). Due to these two system considerations, we will focus on the implementation of VanderLugt correlators throughout the remainder of this chapter.

In the VanderLugt correlator, shown in Figure 5.2, the input image $(f(x, y))$ and the filter function $S^*(f_x, f_y)$ are encoded onto two EASLMs. The input and filter plane SLMs can be operated in transmission-mode or reflection-mode. Reflection-mode devices allow the system to be made more compact, as shown in Figure 5.3, where beamsplitters are employed to accommodate the reflection-mode EASLMs, and to fold the optical path.

The EASLMs can be analog or binary, and modulate the amplitude or phase of an optical beam. Although analog modulators are desirable, it has been shown that binary modulation does not significantly hinder the ability to perform pattern recognition, and can even enhance correlator performance (Psaltis, Paek, & Venkatesh, 1984; Horner & Gianino, 1985; Horner & Leger, 1985; Cottrell, Davis, Schamschula, & Lilly, 1987; Barnes, Matsuda, & Ooyama, 1988). Due to the simplicity of realizing binary EASLMs and the ability to perform correlations with binary inputs and filters, we further focus the discussion onto binary EASLMs and correlator systems employing such devices.

Many algorithms for generating binary Fourier filters have been studied (Psaltis *et al.*, 1984; Horner & Gianino, 1985; Horner & Leger, 1985; Cottrell *et al.*, 1987; Barnes *et al.* 1988; Casasent & Rozzi, 1986;

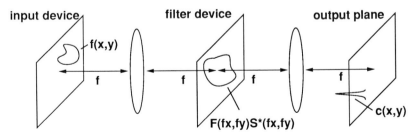

Figure 5.2 Typical VanderLugt correlator layout. The filter plane can be encoded by writing a computer generated filter to some electrically addressable device, or by recording an optical Fourier transform of the filter on an optically addressable device or medium.

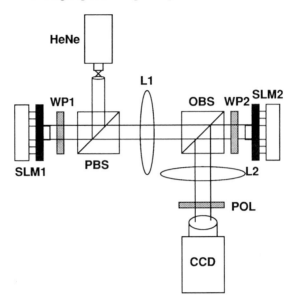

Figure 5.3 Optical correlator system using LCOS SLMs. WP1 and WP2 are half-wave plates, PBS and OBS are polarizing and ordinary beamsplitters, respectively. Lens L1 has a focal length of 200 mm, lens L2 a focal length of 150 mm.

Kallman, 1986, 1987; Flannery, Loomis, & Milkovich, 1988; Jared & Ennis, 1988; Javidi, Odeh & Yu, 1986; Davis, Cottrell, Bach & Lilly, 1989; Downie, Hine, & Reid, 1992; Vijaya Kumar & Bahri, 1989; Leclerc, Sheng, & Arsenault, 1989; Juday, Vijaya Kumar, & Karivaratha Rajan, 1991). One class of binary filters is referred to as binary phase-only filters (BPOFs) (Psaltis *et al.*, 1984). In one particular algorithm for generating BPOFs, the transmittance of the binarized filter made from $s(x, y)$ is set equal to 1 if

$$\mathrm{Re}\,\{S(f_x, f_y)\} > 0 \tag{5.7}$$

and to -1 if

$$\mathrm{Re}\,\{S(f_x, f_y)\} < 0. \tag{5.8}$$

Values of $S(f_x, f_y)$ that lie on the imaginary axis (i.e., $\mathrm{Re}\{S(f_x, f_y)\} = 0$) are set to 1 for

$$\mathrm{Im}\,\{S(f_x, f_y)\} > 0 \tag{5.9}$$

and to -1 for

$$\mathrm{Im}\,\{S(f_x, f_y)\} \le 0. \tag{5.10}$$

Amplitude transmittances of ± 1 correspond to phase shifts of 0 and π and thus the filter generated is called a BPOF. We now review EASLMs suitable for VanderLugt optical corrrelators.

5.2 Electrically addressed SLMs for optical correlation

5.2.1 Introduction

The invention of the liquid crystal light valve (LCLV) started a revolution in optical processing (Grinberg *et al.*, 1975). The LCLV is a programmable, two-dimensional mask that can replace photographic film in optical correlators, holographic systems, and optical interconnection networks, allowing these architectures to operate in 'real-time'. Viable real-time operation of previously static optical processing systems brought about the evolution of other methods of spatial light modulation, including electro-optic, magneto-optic, acousto-optic, mechanical, photorefractive, thermal, and absorptive modulation of the phase, amplitude, and/or polarization of an incident two-dimensional optical wavefront (Efron, 1989).

The first generation SLMs contained pixels generally comprising a modulating material sandwiched between overlapping electrodes, resulting in an EASLM. These successful devices include the nematic liquid crystal-based Hughes Liquid Crystal Light Valve (Bleha *et al.*, 1978) and liquid crystal televisions (LCTV) (Liu, Davis & Lilly, 1985), the magneto-optic-based LIGHT-MOD (Ross, Psaltis & Anderson, 1983), and the ferroelectric liquid crystal (FLC) SLM (STC Technology). Although these EASLMs have been used in proof-of-principle optical processors, their operating characteristics limit their use in high performance optical correlation applications due to their low frame rates (< 1 kHz due to the slow response of the nematic liquid crystal), the high power dissipation and low throughput of the magneto-optic materials, and the passive matrix addressing scheme of the FLC SLMs), large pixel pitch (> 50 μm; yield becomes a problem with narrow electrodes made of thin transparent conductors), and the need for external electronic addressing circuitry which adds complexity, cost, and size to the system.

Recently, a second generation of EASLMs which incorporate electronic circuitry at each pixel has been developed. The second generation devices appear to overcome many of the shortcomings of the first generation devices. These EASLMs have demonstrated high frame rates (> 1 kHz, since an entire array can be electronically programmed before the modulators switch into the desired state), small pixel pitch ($< 30 \mu m$), ease of integration with external drive electronics and controlling computers, low weight, and small volume. In this section, we review EASLM technologies that employ active silicon circuitry at each pixel. These include the lead lanthanum zirconium titanate (PLZT) and silicon (Si/PLZT), deformable mirror device (DMD), and liquid crystal on silicon (LCOS) EASLMs.

5.2.2 Si/PLZT smart SLMs

Introduction

The integration of silicon and PLZT was motivated by the need for faster and 'smarter' SLMs for applications to digital and analog optical processing and optical interconnects (Lee, Esener, Title, & Drabik, 1986). The PLZT modulator is a binary switching element that modulates the phase, amplitude or polarization of an incident optical beam via the transverse electo-optic effect. For a 9–10% doping of lanthanum in the PLZT composition, the modulation is a quadratic function of the applied electric field. For lower or higher lanthanum doping, the modulation is a linear function of the applied field. Generally, the PLZT quadratic electro-optic effect device is preferred in applications to optical processing, since the dielectric constant ϵ_q is four times higher than the ϵ_l obtained with a 12% lanthanum composition (Lee *et al.*, 1986). In this section we describe the functionality of optically and electrically addressed Si/PLZT pixels, and the state-of-the-art in fabricating large arrays of these devices.

Device operation

A typical Si/PLZT functional cell consists of a silicon photodetector, an amplifier, and a PLZT electro-optic modulator (Lee *et al.*, 1986). Electrically addressed arrays have also been fabricated that are directly driven with voltages provided to the chip from external addressing circuitry (Kirkby *et al.*, 1990). The applied voltage appears across two transverse electrodes, modulating the extraordinary index of refraction in the PLZT material. If the incident polarization of the optical field is aligned with the direction of the applied electric field, the phase of the incident beam is modulated. If the incident optical field is polarized at a 45° angle to the applied electric field, and the thickness of the modulator is half-wave in transmission or quarter-wave in reflection, the polarization of the incident optical beam will be rotated by 90°. If an output polarizer analyzes the direction of the polarized light, amplitude modulation or phase modulation can be accomplished.

A Si/PLZT integrated modulator is made by laser recrystallization of Si that has been previously deposited on a polycrystalline PLZT substrate via low pressure chemical vapor deposition (LPCVD). To avoid damaging the PLZT layer, a protective SiO_2 layer is deposited first on the modulator by plasma-enhanced CVD. Transverse, channel electrodes are then etched into the PLZT modulator to produce larger and more uniform transverse electric fields in the PLZT (Lee et al., 1986).

One of the key advantages of the Si/PLZT technology is the speed of optical modulation. Operation speeds of greater than 100 MHz have been achieved in single modulators (Esener, 1992). Arrays of Si/PLZT devices will have a frame rate limited by the ability of the SLM to dissipate power generated by switching the PLZT modulators. Calculations of such frame rate limitations have been carried out and are reviewed below (Lee et al., 1986).

Modeling the PLZT modulators as capacitors, the switching energy dissipation, E_{dis}, for half-wave modulation is given by,

$$E_{dis} = \frac{1}{2} CV_\pi^2, \tag{5.11}$$

with

$$C = \frac{\alpha \epsilon_q wl}{d}, \tag{5.12}$$

where w, l and d are the width, length, and spacing of the pixel electrodes, α is a geometrical factor for the effective cell capacitance, ϵ_q is the dielectric constant, and V_π is the half-wave voltage applied to the electrodes. V_π is a function of the wavelength (λ), the index of refraction (n), the electrooptic coefficient (R), the number of passes through the modulator, m (with $m = 1$ for transmission mode and $m = 2$ for reflection mode), and the modulator dimensions:

$$V_\pi = \sqrt{\left(\frac{\lambda d^2}{n^3 mlR} \right)}. \tag{5.13}$$

It is estimated that the maximum allowable power dissipation in a reflection-mode device (which can be cooled on the backside of the array) is 10 W/cm² (Johnson, McKnight, & Underwood, 1993). Applying typical values ($\alpha = 2$, $\epsilon_q = 5700$, $w = 100\ \mu m$, $l = 40\ \mu m$,

$d = 50 \ \mu\text{m}$, $\lambda = 1.55 \ \mu\text{m}$, $n = 2.5$, $R = 3.8 \times \ 10^{-16}$) to Eqs. (5.11), (5.12), and (5.13) yields the thermal-dissipation-limited operating speed of the modulator to $f_{d_{\max}} = 70 \ \text{kHz}$.

Si/PLZT SLM development

The first Si/PLZT two-dimensional modulator arrays consisted of 16×16 pixels, located on $100 \ \mu\text{m}$ centers (Kirkby, Goodwin, & Parsons, 1990). An 8×8 array of electrically addressable pixels has also been demonstrated. The difficulty in scaling these devices to high space-bandwidth product arrays (greater than 128×128 SLMs) is due in part to the complexity of fabricating the device structure and in part to device power dissipation. This technology shows promise for application to optical processing systems that require a small number of processors (less than 32×32 pixels) and that operate at greater than 10 MHz frame rates, such as optical interconnect networks (Krishnamoorthy, Yayla, & Esener, 1992). Due to the intricate device fabrication and high switching energies required in the Si/PLZT modulator, however, these devices will find less utility for large image processing problems (greater than 128×128 pixels per image), such as optical image correlation. The DMDs and LCOS arrays described next are a better match of the optoelectronic technology to the system application.

5.2.3 DMD SLMs

Introduction

The DMD is an SLM employing electrostatic forces to deform or tilt mirrors located at each pixel. The required electrostatic forces are applied by means of integrated control circuitry on the chip. Mechanical movement or deformation of the pixel mirror due to applied voltages results in a change in the direction or optical path of an input beam that reflects off of the mirror. This path or direction change allows amplitude, phase, or combined modulation of the optical beam (Hornbeck, 1989; Boysel, 1991). The different types of DMD and their use in systems are discussed below.

Device operation

The three types of DMD are based on metalized elastomers, metalized polymer membranes, or metal mirrors mounted on torsion-beam or cantilever-beam structures. All three of these allow the reflective mirror to be attracted to or repelled from a control electrode by controlling the electrode and mirror voltages.

Elastomers are materials that expand (high field) or contract (low field) under the influence of an applied electric field. Deposition of a metalized elastomer material over a control electrode allows deformation of the reflective elastomer surface, and thus, optical modulation (Lakatos & Bergen, 1977). Elastomer-type DMDs are more interesting for display applications than optical processing applications because they operate at slower speeds (~1 ms switching speeds). In addition, elastomers require high voltages (~100–200 V) for switching which makes integration with ordinary silicon circuitry difficult.

A second type of DMD utilizes a metalized polymer membrane, which is stretched over the circuitry and pixel structure. At each pixel a support boundary suspends the membrane over an air gap. An electrode at the bottom of the air gap allows an electric field to be applied across the air gap thereby applying a force to the metalized membrane. The ratio of membrane displacement to air gap thickness can be written as

$$\Delta_0/t = V^2 R^2/T_0 t^3 \qquad (5.14)$$

where Δ_0 is the displacement, t is the air gap thickness, V is the applied voltage, R is the radius of the pixel support boundary (effectively, the pixel size) and T_0 is the membrane tension (Pape & Hornbeck, 1983). This model is only approximate but indicates that the modulation is nonlinear both in applied voltage and in pixel size/shape. Membrane modulation is fast, with turn-on and turn-off times of ~ 25 μs but it also requires relatively high voltages (~ 30 V). The primary disadvantages of the membrane-type DMD are the sensitivity to humidity, pressure, and temperature and the sensitivity/ susceptibility to defects between the membrane and the chip. These disadvantages combined with the aforementioned nonlinear behavior and higher voltage requirements have led to the emphasis on cantilever-beam- and torsion-beam-type DMDs.

The cantilever-beam and torsion-beam DMDs consist of a metal mirror or 'beam' that is attached to a solid support by a thin, flexible piece of metal ('hinge'). Cantilever- and torsion-type DMDs differ from one another in that torsion-type devices use a rotation of the mirror about the hinge axis and cantilever devices use a bending of the hinge to accomplish the mirror movement. In general, torsion-beam structures are better suited to amplitude modulation, while cantilever-beam structures are more appropriate for phase or combined modulation. The single greatest advantage of the cantilever-beam and torsion-beam modulators is their monolithic fabrication. The DMD structures can be fabricated directly on top of processed silicon circuitry, adding only four layers to the entire fabrication process. After fabrication of the control circuitry (for example, in an ordinary complementary metal oxide semiconductor (CMOS) process) the independent DMD process is begun. The four layers are the electrode, spacer, hinge, and beam layers. A combination of patterning, etching, and oxide deposition involving these four layers forms the structure of the DMD, and after the wafers are cut into individual chips, a final etch removes the material underneath the mirrors and hinges, leaving the completed DMD structure.

Detailed analysis of the relationship between mirror deflection, applied voltage, and response time for torsion-beam and cantilever-beam DMDs is beyond the scope of this chapter, but can be found in Hornbeck (1989). The torsion and cantilever-type DMDs can operate in both analog and binary modes and require low power for switching. Response times as low as 10 μs are obtainable using moderate applied voltages (5–20 V).

Device development

A 128 × 128 metalized polymer DMD has been fabricated and tested (Hornbeck, 1983; Pape & Hornbeck, 1983). The device has pixels located on 51 μm centers and rectangular air gaps that are 23 μm × 36 μm. The average contrast ratio of this device is 10:1 as measured over an entire pixel area.

More recently, a wide variety of torsion-beam and cantilever-beam DMDs have been demonstrated and used in optical processing systems. Such devices range from a 128 × 128 frame-addressed device (Boysel, 1991) to a 768 × 576 display device (Sampsell, 1993). Optical correlation has been performed using a 128 × 128 line-addressed DMD as a Fourier filter in a joint transform architecture (Florence & Gale, 1988; Florence, 1989). Pixel spacings on these devices are 50 μm and a typical contrast ratio is 10:1. The DMD SLMs used in these systems are electrically addressed, and thus the systems employ a static input or an intermediate detection stage to operate as a JTC. This necessity limits the otherwise high DMD frame rates ($\sim$ 8 kHz for frame-addressed DMDs) to video rate system operation for nonstatic inputs. Use of a VanderLugt architecture would eliminate this speed bottleneck and also improve the space-bandwidth capacity of the system. DMD SLMs have demonstrated their viability in folded spectrum analyzers (Boysell, 1991), printing (Nelson & Hornbeck, 1988), neural networks (Collins, Sempsell, Florence, Penz, & Gately, 1988), and display applications (Sampsell, 1993). Work is in progress on the development of high resolution 2048×1152 pixel DMD arrays for high definition television (HDTV) applications (Sampsell, 1993).

5.2.4 Liquid crystal on silicon EASLMs

Introduction

Nematic and cholesteric liquid crystals have been utilized as electro-optic materials in a number of applications requiring low drive voltages, large apertures, high resolution, and low cost. Their principal application is in information display; watches, calculators, and televisions. They operate on the basic principle that due to their large dielectric anisotropy the average molecular axis, or director, can be aligned in the direction of an applied electric field by an electrically induced torque. The torque is a quadratic function of the electric field strength, E. Even though the molecules can align to the electric field in < 1 ms, the molecules relax back to their original state through a mechanical process, limiting the response times of nematic liquid crystals to tens of milliseconds (DeGennes, 1975). As the nematic liquid crystal is cooled, the molecules may self-assemble into layers. The layered phases are known as smectics. If the long axis of the rod-shaped molecule is perpendicular to the layers, the phase is called smectic A (SmA). Lowering the temperature of the liquid crystal further may induce a tilt in the molecules with respect to the layer normal, forming the smectic C (SmC) phase. The symmetry of the SmC phase is C_{2h}, which means there is two-fold rotational symmetry about an axis perpendicular to the molecular director and parallel to the smectic layers, and a mirror plane congruent with a plane containing the layer normal, z, and the molecular director (known as the tilt plane).

Meyer recognized in 1974 that the mirror symmetry disappears if the liquid crystals are chiral, reducing the symmetry of the SmC to the C_2 polar point group (Meyer, Liebert, Strzelecki, & Keller, 1975). He further postulated the existence of ferroelectricity in chiral smectic liquid crystals (CSLC), meaning the existence of a permanent dipole moment that can be switched by an applied electric field.

Clark & Lagerwall (1984) fabricated the first bulk, ferroelectric liquid crystal switch by using bounding substrates to surface stabilize thin layers (< 5 μm) and thus remove the helix. The smectic planes are perpendicular to the glass plates in the so-called 'bookshelf' geometry. The result is a device with two approximately equal energy states oriented at $\pm\ \psi$ to the layer normal, as shown in Figure 5.4. The consequence of this simple solution to creating macroscopic polarization is the ability to couple directly an applied electric field, E, to the spontaneous polarization, P_s. Since molecular alignment is accomplished by the torque, $P_s \times E$, reversing the sign of E reverses the sign of the torque causing molecular reorientation in the opposite direction. Clark & Lagerwall (1980) reported submicrosecond switching in the FLC HOBACPC at 68 °C with approximately 30 V/μm applied electric field.

Applications involving FLCs take advantage of their large birefringence, Δn, by rotating the molecular directors in a plane parallel to the glass substrates. Typical values for Δn range from 0.1 to 0.2. When the thickness of the material, d, satisfies the equation,

$$d = \lambda/2\Delta n, \tag{5.15}$$

where λ is the wavelength of light in vacuo, a π phase shift occurs between the ordinary and extraordinary waves of light propagating through the birefringent material. The device is essentially a switchable half-wave plate. When $\lambda = 0.632$ μm and $\Delta n = 0.15$, the zero-order half-wave thickness $d = 2$ μm.

An optical intensity modulator is made by placing the CSLC half-wave plate between crossed polarizers. When the material optic axis is aligned parallel or perpendicular to the incident polarizer, no rotation

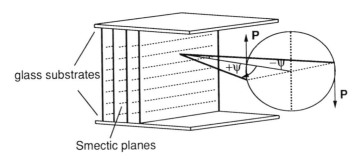

Figure 5.4 Surface stabilized FLC structure. Molecules can lie at either $+\psi$ or $-\psi$ to the smectic layer normal. **P** represents the spontaneous polarization direction.

of polarization occurs, and the light is extinguished by the output analyzer. Reversing the sign of E rotates the optic axis by 2ψ, resulting in a rotation of incident linearly polarized light by 4ψ. When $\psi = 22.5°$, incident light is maximally transmitted by the output analyzer.

Binary phase modulation is accomplished with FLC half-wave plates by orienting the input polarizer along the bisector of the two maximum switching states and the output polarizer perpendicular to the input polarizer. The maximum throughput occurs when $\psi = 45°$. In the following, we describe the ferroelectric LCOS EALSMs used as both binary amplitude and phase modulators in the optical correlators described in Section 5.4. These devices were designed to accept digital electronic signals and to produce binary optical outputs.

Device operation

LCOS SLMs are the result of the merging of advanced liquid crystal and very-large-scale integrated (VLSI) circuit technologies. The silicon VLSI circuitry controls the local voltages across a liquid crystal modulating layer, as illustrated in Figure 5.5. This allows a spatial variation of the liquid crystal state and thus produces an SLM.

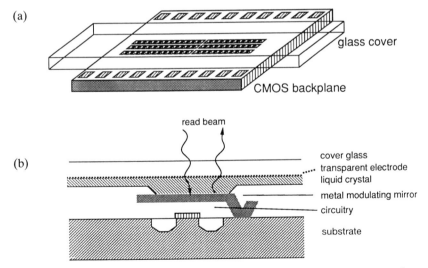

Figure 5.5 (a) LCOS SLM. Liquid crystal is sandwiched between cover glass and a silicon backplane. (b) Cross-section of a LCOS SLM. Cover glass is coated with a transparent conductor allowing for a controlled electric field between the metal modulating mirror (controlled by the on-chip circuitry) and the cover electrode.

This technology was first applied to the manufacture of miniature display devices (Crossland & Ayliffe, 1981; Shields, Fletcher & Bleha, 1983). The flexibility of silicon circuit design and the speed of advanced liquid crystal mixtures led researchers to consider a variety of devices for optical processing applications (Underwood, Vass & Sillitto, 1986; McKnight, Vass & Sillitto, 1989; Cotter, Drabik, Dillon & Handschy, 1990; Jared & Turner, 1991; Jared, Turner, & Johnson, 1991). For the optical correlator applications discussed in this chapter, two different pixel and array designs have been implemented and are discussed below.

LCOS pixel designs used in optical correlators

The one transistor dynamic random access memory (DRAM) The one transistor pixel circuit is shown in Figure 5.6. The gate of the transistor is connected to the row address line, the drain to the column data line, and the source to the metal modulating mirror, used to apply a voltage to the liquid crystal modulating pad. The advantages of this approach to pixel design include low circuit complexity and, consequently, small pixel size. The primary disadvantages are the limited amount of charge available to switch the liquid crystal (if the gates are clocked quickly compared to the response time of the liquid crystal), and the need for data to be refreshed from a framestore (even when it is not being altered). Furthermore, the rate of charge leakage from capacitive nodes is increased in the presence of light. However, the intrinsic bistability of the surface stabilized FLC may make this a moot point (Xue, Handschy & Clark, 1987).

The four transistor DRAM The object of this pixel style is to decouple the 'soft' capacitive node storing the state of the pixel from the node

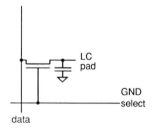

Figure 5.6 One transistor DRAM pixel element. When *select* is HIGH *data* are transferred to the capacitance of the liquid crystal modulating pad.

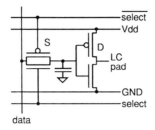

data

Figure 5.7 Four transistor DRAM pixel element. When *select* is HIGH and *select* is LOW *data* are transferred to the input capacitance of the inverter, driving the liquid crystal modulating pad.

which drives the FLC by means of an inverting buffer as shown in Figure 5.7. In principle, this allows an unlimited amount of charge to drive the FLC which removes one of the drawbacks of the one transistor DRAM, and allows the use of higher P_s FLC mixtures for faster switching.

Both the four transistor and one transistor pixel designs can employ metal shielding to reduce degradation of circuit performance due to the incident illumination. The drawback to this shielding technique is a reduction of the flat pixel area.

LCOS EASLM development

64 × 64 EASLM array The optical correlator described in Section 5.4 employs LCOS EASLMs with 64 × 64 pixels located on 40 μm centers (Jared *et al.*, 1991). This array uses the four transistor DRAM pixel design, with the metal modulating mirror overlapping the transistor circuitry. A 28 × 28 μm cut in the glass passivation layer yields an effective fill factor of 49%. Sixteen parallel lines feeding a 16 × 64 multiplexer were used to address the array columns. Row addressing was accomplished by a 1 × 64 dynamic shift register.

The backplane was fabricated in the 2 μm, *n*-well CMOS process by VLSI Technologies through the NSF/ARPA supported University of Southern California Metal Oxide Semiconductor Implementation Service (MOSIS) foundry. After processing, the die are packaged into 84 pin grid array packages. A thin layer (~ 3 μm thick) of BDH SCE13 smectic C* FLC is sandwiched between the packaged silicon backplane and a 5 mm × 6 mm piece of glass cut from a BK7 optical flat. The glass substrate, which is coated with a transparent conductive oxide (TCO) electrode and a rubbed polyvinyl alcohol alignment layer, is epoxied to the chip. Polyball spacers 1.7 μm thick provide a cavity for the liquid crystal to flow onto the silicon substrate. A wire attached to the top electrode provides a method for controlling the electric field between the chip metal modulating pads and the TCO. Figure 5.8 shows a photograph of the packaged SLM.

For optimal modulation efficiency, the FLC layer thickness that results in a low-order, half-wave retarder for the desired operating wavelength is preferred, since the retarder is more achromatic at lower waveplate orders. Furthermore, the electric field strength across the FLC layer decreases as cell thickness increases, which results in a longer FLC switching time, τ_{lc} (Xue *et al.*, 1987). Although a zero-order half-wave retarder is desirable, it is currently difficult to obtain a uniform gap at the specified thickness across the chip. The filling procedure used to fabricate these EASLMs used spacers which resulted in first-order FLC waveplates at $\lambda = 632.8$ nm (~ 3 μm FLC layer).

Binary intensity modulation of light is achieved by applying two equal amplitude, but opposite polarity voltages across the FLC. This is accomplished by holding the top electrode at 2.5 V, which for ordinary CMOS circuitry is half-way between the 0 and 5 V available at the modulating pad. This switches the optic axis of the FLC between two orientations, separated by approximately 45° for smectic C* FLC materials. For the optimum wavelength satisfying Equation (5.15), the device functions like a programmable half-wave plate. Measurements taken on this first device showed a ~ 300 μs liquid crystal response

Figure 5.8 Photograph of LCOS EASLMs.

time (τ_{lc}), and a contrast ratio of 5:1 over a 16 × 64 block of pixels in the array. The shift registers operate at over 16 MHz, corresponding to an array electronic addressing time of 16 μs (τ_{elec}). This puts achievable frame rates for this device, determined by the sum of $\tau_{lc} + \tau_{elec}$, at greater than 3 kHz.

128 × 128 EASLM array The EASLM used in the correlator described in Section 5.4.2 is a 128 × 128 pixel array with the pixels located on 30 μm centers (Johnson, McKnight, & Underwood, 1993). The data are loaded onto the device through sixteen parallel input lines each driving an eight-bit static shift register. The rows are selected by passing a token down two sets of 1 × 64 static shift registers. The column and row shift registers are located on 60 μm centers, and interleaved to address the 30 μm-centered pixels. The chip was fabricated through the MOSIS 2 μm, n-well process at VLSI Technologies.

EASLMs were assembled from the processed die in the same manner as the 64 × 64 four transistor EASLMs. The pixel modulating mirror is designed to minimize liquid crystal thickness variations within the active area of the array. All of the circuitry at a pixel and all address lines to a pixel are to the side of the mirror. The modulating mirror is approximately 26 μm × 26 μm, with a 22 μm × 22 μm overglass cut yielding a maximum fill factor of 54%. However, measurement of the chip surface indicates the actual overglass cut limits the fill factor to 30%. Unpassivated die would allow a maximum fill factor (design-rule-limited) of nearly 70%.

The optical throughput of the SLMs was measured to be a maximum of 19%, which is a function of the fill factor (30%) and the modulating mirror reflectivity (40–60%). The measured SLM liquid crystal response time is $\tau_{lc} = 150$ μs. The shift register clock speed is 20 MHz and with 16 parallel input lines the minimum electronic load time, τ_{elec}, is 51 μs. Hence the frame time, $\tau_{elec} + \tau_{lc}$, corresponds to a maximum SLM frame rate of 5 kHz. Figure 5.9 is a photograph of the SLM displaying a checker-board pattern. Each block consists of an 8 × 8 group of pixels. The contrast ratio is approximately 18:1, as estimated from measurements made on a row of pixels. The SLM from which these measurements were made is a first-order half-wave plate at $\lambda = 510$ nm.

Thus far, we have reviewed SLM technology that is suitable for high speed optical correlator applications. Because optical correlators with active input and filter plane devices that are both high speed and

Figure 5.9 128×128 binary LCOS EASLM displaying a checker-board pattern.

compact have been primarily based on liquid crystal technology, we shall now focus on correlator design based on LCOS technology. The LCOS EASLM device characteristics having the greatest influence on correlator performance include device size, pixel size, device contrast ratio, backplane flatness and uniformity, pixel fill factor, and speed. In the following sections each of these characteristics and their influences on optical correlator performance are discussed.

5.3 Influence of LCOS EASLM device characteristics on correlator performance

5.3.1 Pixel and device size

The effect of pixel pitch, size, and shape on the performance of an optical correlator is related to the Fourier spectrum of the individual pixel. Consider the Fourier transform of an image, $f(x, y)$, encoded on an $N \times N$ array of square pixels of dimension a located on centers spaced by a distance d_{pix}. The encoded image, $f_{\mathrm{slm}}(x, y)$ can be written as the convolution of a rectangle function of square dimension a with the sampled input $f(x, y)$,

$$f_{\mathrm{slm}}(x, y) = \left[f(x, y) \,\mathrm{comb}\left(\frac{x}{d_{\mathrm{pix}}}, \frac{y}{d_{\mathrm{pix}}} \right) * \mathrm{rect}\left(\frac{x}{a} \right) \mathrm{rect}\left(\frac{y}{a} \right) \right] \mathrm{rect}$$
$$\left(\frac{x}{Nd_{\mathrm{pix}}} \right) \mathrm{rect}\left(\frac{y}{Nd_{\mathrm{pix}}} \right), \tag{5.16}$$

where the comb function (Goodman, 1968) sample spacing is the pixel pitch, d_{pix}. The Fourier transform of Eq. (5.16) is given by

$$\mathcal{FT}(f_{\mathrm{slm}}(x, y)) = N^2 d_{\mathrm{pix}}^4 a^2 [[F(f_x, f_y) * \mathrm{comb}(f_x d_{\mathrm{pix}}, f_y d_{\mathrm{pix}})] \,\mathrm{sinc}$$
$$(f_x a)\,\mathrm{sinc}(f_y a) * \mathrm{sinc}(f_x Nd_{\mathrm{pix}})\,\mathrm{sinc}(f_y Nd_{\mathrm{pix}})]. \tag{5.17}$$

The pixel pitch and shape determine the envelope function that modulates the image spectrum, which for flat, square pixels is sinc $(f_x a)\,\mathrm{sinc}(f_y a)$. This envelope modulates the information at both the filter and correlation planes. Smaller pixels generally correspond to a higher space-bandwidth product and more slowly varying envelope functions and are therefore less likely to introduce aliasing artifacts and other errors. On the other hand, larger pixels reflect or transmit more optical energy through the system and reduce dead space. Thus to maximize system throughput it is desirable to increase the pixel area. These two competing effects find a compromise in actual hardware implementations.

Technology-based constraints on array size, pixel pitch, and pixel size determine the minimum possible pixel spacing. In the case of LCOS modulators, pixel pitches range from 20 μm to 100 μm. Linear pixel dimension may be anywhere from 70–90% of the center-to-center pixel pitch, d_{pix}. Based on the Nyquist sampling criteria the pixel pitch determines the maximum spatial frequency, f_{Max}, that the array can encode,

$$f_{\mathrm{Max}} = 1/2d_{\mathrm{pix}}. \tag{5.18}$$

From Eq. (5.18) we can express the total number of pixels as the product of the spatial frequency bandwidth and the device area, A,

$$N^2 = 4Af_{\mathrm{Max}}^2. \tag{5.19}$$

This quantity is referred to as the space-bandwidth product (SBWP) and is a measure of the information capacity of the SLM.

With respect to the filter plane, the pixel pitch and SBWP have a direct influence on the correlator pattern recognition performance. A filter consisting of discrete pixels results in multiple replications of the desired correlation function at the output plane of the optical system. Aliasing between these replications can cause the correlation to be degraded or completely masked. Manipulating the pixel spacing and profile can be used to influence the strength of the replicated orders in systems where aliasing is a concern (Gheen & Washwell, 1990; Heddle, Vass & Sillitto, 1992).

The aliasing problem can be avoided by using a high bandwidth filter or by blocking the aliased frequencies at the filter plane. The SBWP of the filter device determines the level of detail of the filters encoded on the filter SLM and increased filter SBWP thus improves the ability of the system to recognize different objects.

The overall system length of a VanderLugt correlator is also influenced by the SLM pixel pitch. The extent of the optical Fourier transform scales linearly with the focal length of the transform lens and the wavelength of the illumination,

$$f_x = x/\lambda f, \, f_y = y/\lambda f \qquad (5.20)$$

where f_x and f_y are spatial frequency variables, x and y are physical distances, f is the focal length of the transform lens, and λ is the wavelength of illumination. Setting $f_x = f_{\text{Max}}$ fixes the physical position of the maximum frequency component in the Fourier domain and thus sets the minimum size of the filter plane array if no information is to be lost. Ideally, the filter plane device should be no larger than $\lambda f f_{\text{Max}} \times \lambda f f_{\text{Max}}$ to prevent transmission of excess high frequency noise. The condition where the Fourier transform of the input device is exactly matched to the size of the filter plane device is referred to as 1:1 scaling. The focal length required for 1:1 scaling between the input and the filter is given by

$$f = N d_{\text{pix}_i} d_{\text{pix}}/\lambda, \qquad (5.21)$$

where d_{pix} and N are defined in Eqs. (5.18) and (5.19), and the subscript i denotes the input SLM (Davis, Waring, Bach, Lilly, & Cottrell, 1989). Eq. (5.21) shows that correlator system size, determined by f, scales linearly with the number of pixels and quadratically with pixel pitch. Thus, a correlator utilizing SLMs with $d_{\text{pix}} = 100 \, \mu\text{m}$ will require a focal length that is an order of magnitude greater than that required for a correlator with $d_{\text{pix}} = 30 \, \mu\text{m}$ SLMs.

Technology limitations on pixel size and device SBWP have resulted

Table 5.1 *Pixel pitch, fill factor, and 4-f correlator 1:1 scaling focal length ($\lambda = 632.8$ nm) for various SLMs.*

SLMs	Pixel pitch (μm)	Fill factor[a] (%)	Focal length (cm)
Seiko 240 × 640 LCTV	530	85%	10.6×10^4
Semetex 128 × 128 magneto-optic Sightmod	76	54%	117 cm
STC 128 × 128 matrix addressed FLC	165	72%	550 cm
T1 128 × 128 deformable mirror	25	80%	13 cm
Displaytech/Georgia Inst. of Tech. 64 × 64 LCOS	60	56%	36 cm
Univ. of Colorado 64 × 64 LCOS	40	49%	16 cm
Boulder Nonlinear Systems 128 × 128 LCOS	30	54%	18 cm
Displaytech 256 × 256 LCOS	20	70%	16 cm
STC 176 × 176 LCOS	30	34%	25 cm
Univ. of Colorado 256 × 256 LCOS	21	79%	16 cm

[a]Fill factor here is quoted as area of pixel divided by pixel pitch. Mirror flatness is neglected here, but is discussed in Section 5.3.4.

in clever optical designs to reduce the system size. Davis *et al.* (1989) and Flannery, Biernacki, Loomis, & Cartwright (1986), have designed correlators which reduce the size of a $4f$ correlator by more than a factor of 10. To accomplish the Fourier transform operation properly such designs increase the number of lenses and requirements on their optical quality. The complexity and cost of such optical correlators is therefore also increased.

Referring again to Eq. (5.21) we find that reduction of the pixel size on the SLM results in a smaller optical system with the same processing power. Silicon technology is well suited to accomplishing this type of reduction. Table (5.1) compares pixel spacings and required focal lengths for 1:1 scaling (from Eq. (5.21)) for various commercially available EASLMs and existing LCOS devices (Displaytech, Inc; Thorn EMI; Jared *et al.*, 1991; McKnight, Johnson, & Serati, 1994).

5.3.2 Pixel fill factor

We define the fill factor as the pixel area divided by the pixel pitch squared. In terms of pixel dimension a and pixel pitch d_{pix}, the pixel fill factor, β, is given by

$$\beta = a^2/d_{\text{pix}}^2. \qquad (5.22)$$

Control circuitry required to set the state of the pixels reduces β. As EASLM performance requirements increase, silicon design rules for spacings between the different layers, such as metals, polysilicon, and diffusions, place a greater restriction on β.

There are two dominant effects of reduced fill factor on correlator performance. The first effect concerns system energy efficiency. Pixels, whether reflective or transmissive, will transmit light through the system based on their modulating area. This is verified by examining Eq. (5.17) which expresses the amplitude of the Fourier transform of a function encoded onto an SLM with discrete pixels. The scaling factor in front, $N^2 d_{\text{pix}}^4 a^2$, has the fill factor embedded in it and can be expressed in terms of β as $A d_{\text{pix}}^4 \beta$, where A is the array area. If the device area and pixel spacing are fixed, the amplitude of the Fourier transform scales linearly with β, and the power spectrum of light modulated by the EASLM scales as β^2. The University of Colorado 64 × 64 LCOS modulator (see Table 5.1) has a fill factor of $\beta = 0.49$. This indicates that the Fourier spectrum of light modulated by this device experiences a 75% power loss due to reduced fill factor. Further reflection from a second EASLM, such as in a VanderLugt correlator, results in a 92.8% loss of optical power. Such a reduction must be considered when choosing illumination sources and detection devices.

The second major effect of reduced fill factor is system noise generated by light that reflects off or is transmitted through the nonpixel areas of a modulator. This effect results in an increased zero-order component, combined with light that is diffracted by structures in the nonpixel areas. The magnitude of the increased zero-order and diffracted light scales in intensity with β^2 and is also related to the transmittance/reflectance of the dead space. We can see the effects on the phase-only filtering operation by comparing the zero-order strength for different values of β. For a 100% fill factor device ($\beta = 1$) the intensity of the zero order is 0 for a BPOF generated from a simulated square aperture. If β is reduced to 0.5 and the nonpixel areas are assigned a fixed modulation, the zero order for this filter increases to 26% of the total power. Similar results can be obtained for filters that have arbitrary frequency content. For the total correlation operation, the fill factor effect has been analyzed assuming 0 transmittance in the dead areas between pixels (Gianino & Woods, 1992). The results of

Gianino & Woods' analysis indicate that other than a reduction in optical throughput, correlator performance is not significantly influenced by fill factor. In the case of nonopaque space between pixels, measures of correlator SNR that are sensitive to zero-order level (Vijaya Kumar & Hassebrook, 1990; Horner, 1992) will be affected by the dead zone area, but the detectability of the peak should not be significantly reduced. The primary effect will be the reduction in throughput that scales as the fourth power of β.

5.3.3 Contrast ratio

The *off* state of LCOS SLMs will be nonzero due to liquid crystal defects, scattered light, imperfect polarizers, and stray reflections. This nonideal *off* state combined with a reduced *on* state intensity transmittance causes a reduction in the device contrast ratio. If we consider the frequency content of an input amplitude encoded image, the reduced contrast may result in an increase in the zero-order frequency component. This is most easily seen by considering the extreme case of minimum contrast ratio, $T_{on}/T_{off} = 1$. In this case, all of the diffracted light appears in the zero order of the Fourier transform of the illuminated device.

Reduced contrast ratio and contrast ratio variations influence the filter plane of a correlator differently. Assuming a BPOF, we find that the amplitude of the 0 and π phase shifted components will vary due to contrast ratio effects. Because the information is contained in the phase, reduced contrast ratio, on its own, does not strongly affect the phase-only filtering operation. Simulations of reduced contrast by Horner (1990) indicate that there is little difference in phase-only filter performance when contrast is reduced from 100:1 to 5:1.

Some errors that affect contrast ratio influence the phase of the reflected light, most notably thickness variations in the liquid crystal layer. With these types of error, an apparent relationship between phase-only filter performance and contrast ratio may be inferred. Device defects that result in liquid crystal layer thickness variations can simultaneously cause reduced contrast ratio and poor phase-only filtering operation. Contrast ratio alone does not significantly reduce the performance of the phase-only filter. The cause and effect of device thickness variations which result in reduced correlator performance are examined in the next section.

5.3.4 Backplane flatness and uniformity

The relative phase and the absolute phase introduced by the LCOS modulator depend on the thickness of the liquid crystal layer (Turner, Jared, Sharp & Johnson, 1993). A perfect modulator introduces a constant absolute phase and the desired relative phase determined by the state of the liquid crystal at each pixel. LCOS modulators suffer from liquid crystal thickness variations that cause a deviation from the desired quantities of relative and absolute phase. The primary types of thickness variations across the modulator include warping of the silicon backplane, small pixel-to-pixel height variations of the modulating mirror, and structure within a pixel resulting in intra-pixel variations. The pixel-to-pixel variations (and some backplane warpage) are inherent in the processing of the silicon die.

The backplane appears to be warped when the die is epoxied to the chip carrier. A dual-path Michelson interferometer was used to observe the fringe patterns of the warped backplanes, as shown in Figure 5.10(a) and (b). If the warping is assumed to be spherical, the unpackaged die flatness of 0.7λ over 7.1 mm corresponds to a radius of curvature of 14 m. The flatness after packaging is 2.25λ corresponding to a radius of curvature of 4.5 m.

Variations in the height of the metal modulating mirrors within a pixel cause additional variations in the liquid crystal thickness. Such mirror height variations exist due to integrated circuit structure and semiconductor processing effects. The thickness variations due to circuitry and processing effects that exist within a pixel represent spatial frequencies that are equal to or higher than the maximum system frequency and are therefore not a significant source of system noise. These variations do, however, reduce the system throughput in addition to the previously discussed fill factor effects. Inter-pixel thickness variations add to the system noise and are discussed below.

To understand how the thickness variations influence system performance, these effects were incorporated into a computer model and their impact on performing optical correlation simulated. Investigations on the effect of warping in both magneto-optic (Downie *et al.*, 1992) and LCTV (Casasent & Xia, 1986; Mok, Diep, Liu & Psaltis, 1986; Kim, Jeong, Kang & Jeong, 1988; Horner, 1988) devices treat this source of system noise as an absolute phase error. The thickness variations that exist across the LCOS device alter the thickness of the FLC layer and hence change both the absolute and the relative phase modulation characteristics of the device. Therefore, simulations of

(a)

(b)

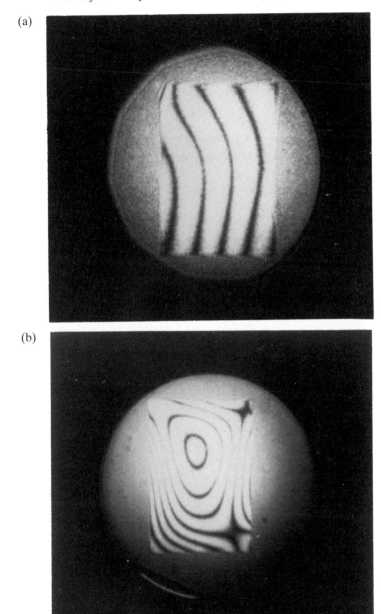

Figure 5.10 (a) Interferogram of unpackaged silicon backplane. Center-to-edge surface variation is 0.7λ (λ = 632.8 nm) corresponding to a radius of curvature of 14 m. (b) Interferogram of packaged backplane. Center-to-edge surface variation is 2.25λ corresponding to a curvature of 4.5 m.

thickness variations in LCOS SLMs model the influence of both the absolute and retardation phase effects on correlator system performance.

In the computer model, pixel-height standard deviation and backplane curvature were used as variable parameters. Although the warping seen on the devices is not perfectly spherical, for the purposes of the computer simulation it was modeled as such. Typical LCOS modulators have radii of curvature between 3.0–5.0 m (Turner et al., 1993). Therefore values of curvature ranging from 0.05–40.0 m were simulated to cover a wide range of values.

The pixel-to-pixel variations were modeled by randomly perturbing the device thickness at each pixel. The perturbations were obtained from a zero-mean Gaussian distribution and added to the modulator thickness at each pixel. Data from the MOSIS service, through which the integrated circuits are processed, suggests that the typical standard deviation for layer thickness variations is 150 Å (University of California Information Sciences Institute). Values of pixel-height standard deviation ranging from 10–1000 Å were simulated. For these simulations, the input to the correlator was a random array of binary values. A BPOF was generated from each input array as discussed in Section 5.1 and the thickness variations were introduced to both the inputs and filters. Autocorrelations for 100 different input/filter pairs were computed for each curvature and pixel-height standard deviation, to minimize data-dependent effects and statistical fluctuations. It is important to note that a real system would be affected by other types of noise in addition to the thickness variations and a great deal of work has been done on the effects of system noise on correlator performance (Javidi & Yu, 1986; Horner & Gianino, 1987; Dickey, Stalker & Mason, 1988).

The effects of the two types of thickness variation on correlator performance are shown in Figure 5.11. The performance measure chosen is the peak-to-correlation energy (PCE) ratio, as defined by Vijaya Kumar & Hassebrook (1990). This measure indicates the percentage of energy that exists in the correlation peak in comparison to the total energy in the correlation plane. For both curvature and pixel-to-pixel variation the data in Figure 5.11 are divided into three regions: a region where there is no correlation, a region where performance changes rapidly with modulator quality, and a region where the performance is maximized and becomes relatively independent of further improvement in the flatness of the modulator.

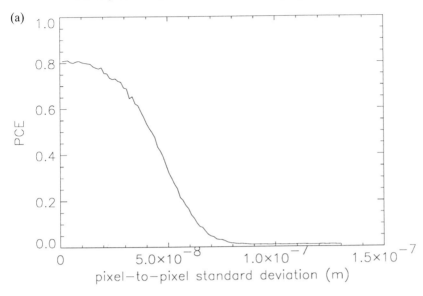

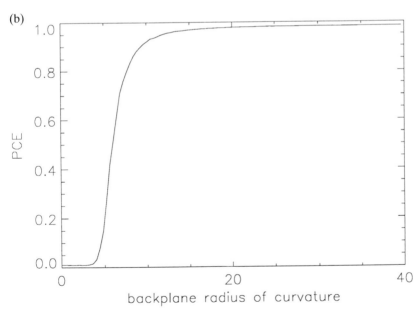

Figure 5.11 (a) Peak-to-correlation energy plotted against standard deviation of pixel-to-pixel height variation. (b) Peak-to-correlation energy plotted against backplane radius of curvature.

Computer simulation results indicate that a pixel height standard deviation of 150 Å, corresponding to typical silicon processing parameters, results in system performance that is within 10% of the maximum noise-free performance. Hence we conclude that pixel-to-pixel height variation is not likely to be a limiting factor in overall correlator performance. This is not the case for backplane curvature.

Figure 5.11(b) shows that for correlator performance at 90% of the maximum PCE, the radius of curvature of the backplane must be 10.0 m or more. This radius of curvature corresponds to a flatness of 1.0λ. Referring to Figure 5.10(a) it is seen that the unpackaged backplane can meet this curvature requirement and that it should be theoretically possible to get PCE performance that is within 90% of the maximum.

The curvature found in packaged devices (typically 3.0–5.0 m) puts system performance at a level that is less than 10% of maximum performance. For example, a device curvature of 4.5 m (see Figure 5.10(b)) indicates a PCE performance of 4.1% of maximum in our computer model. We see that the model predicts correlator system performance that is substantially below optimum due to the backplane warping.

Several approaches have been taken to reduce the amount of backplane warpage. Methods that are currently under investigation include backplane planarization and polishing (O'Hara, Hannah, Underwood, Vass & Holwill, 1993), low stress epoxies (Jared, 1993), and solder bump bonding (Lin, Lee & Johnson, 1993).

5.3.5 Speed

Although the first VanderLugt optical correlator used film for both the input and filter planes, rapid correlation between real-time input scenes and a bank of filters is imperative for practical applications of optical image correlation systems. Pattern recognition can be performed by correlating digitized image information. Dedicated digital signal processing chips can currently accomplish this at speeds ranging from 15–40 frames per second on 256×256 arrays of data (Hose, 1992; Array Microsystems; United Technologies Microelectronics Centre). In the near future such electrically-based correlators are expected to achieve speeds of 200 frames per second (Hose, 1992). For an optical correlator to be feasible, it must be able to meet or exceed these operation speeds while maintaining the inherent optical advantage in SBWP, in addition to passing tests of dynamic range, weight, cost, and volume.

As discussed in Section 5.2.4, the speed or frame time of an electrically addressed LCOS device has two components,

$$\tau_{frame} = \tau_{elec} + \tau_{lc}, \qquad (5.23)$$

where τ_{elec} is the time required to address electrically or load the data onto the SLM and τ_{lc} is the time required for the liquid crystal to switch after it is addressed. Improvements in both the FLC modulator and the electronic circuit design will affect the correlator system speed. Using standard CMOS circuitry, voltages available to switch the liquid crystal are generally limited to less than 7 V. At these voltages, τ_{lc} is the limiting factor in device speed. For existing LCOS devices, τ_{lc} ranges from 150–300 μs and τ_{el} from 10–50 μs (Underwood, Vass & Silletto, 1986; Drabik & Handschy, 1990; Jared *et al.*, 1990). In the future it is reasonable to expect τ_{lc} to be reduced to less than 25 μs using more advanced FLC materials and high voltage circuitry (Johnson *et al.*, 1993). This will allow device frame rates of higher than 10 kHz. If SLMs achieve these speeds, then the correlator output must be detected and processed at similar rates. This issue is addressed in Section 5.5.2.

5.4. Experimental results from optical correlators employing LCOS EASLMs

Thus far two correlators have been implemented employing LCOS EASLMs as both real-time input and filter plane devices. The first employed 64 × 64 EASLMs and the second 128 × 128 EASLMs. A description of these processors and their performance is given below.

5.4.1 64 × 64 LCOS-based optical correlator

An optical correlator system was demonstrated using 64 × 64 LCOS EASLMs as both the input and filter planes (Turner *et al.*, 1993). Figure 5.3 shows the optical system design. A collimated input beam ($f = 75$ mm, $D = 25.4$ mm, collimating lens) from a HeNe laser ($\lambda = 632.8$ nm) is reflected by the polarizing beamsplitter and transmitted through a half-wave plate which properly orients the input polarization relative to the FLC optic axis. Light reflected by SLM1 is

encoded with the input pattern producing a binary amplitude modulated image at the output of the polarizing beamsplitter (PBS). Lens
L1 (f = 200 mm, D = 50 mm) performs the Fourier transformation of
the input, which is multiplied by the filter pattern, encoded as a BPOF
on SLM2. Another half-wave plate precedes SLM2 to reorient the
polarization for the phase-only filtering operation. The resulting wavefront is reflected by the ordinary beamsplitter through lens L2
(f = 150 mm, D = 50 mm) which forms the Fourier transform of the
input/filter product at the camera plane. The y-oriented analyzer
removes the unwanted polarization component. The entire experiment
is controlled by an IBM compatible 386-based personal computer. Two
custom built interface boards drive the SLMs, and the Cohu 4815-2000
CCD camera output is stored in the computer using an Imagenation
Cortex-I frame grabber.

The input pattern used to test the optical correlator was a circle
displayed on SLM1. Since the LCOS EASLM's operating frame rates
exceed those of the CCD camera, the same input and filter data were
written to the devices continuously to allow the CCD camera to
capture the correlation data. Data from the camera were digitized and
converted to the surface plot shown in Figure 5.12. The value of PCE
is 3.68%, as indicated in the plot and the peak-to-secondary ratio,
defined as the peak height over the height of highest sidelobe, is 1.85.

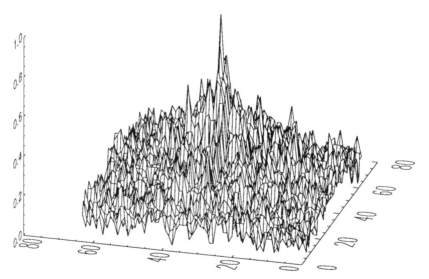

Figure 5.12 LCOS correlator output with PCE = 3.68%.

Referring back to Figure 5.11 we see that the value of PCE is consistent with the value predicted by the simulations for the amount of backplane warpage that exists in current devices. From these results it appears that the curvature of the silicon backplane significantly reduces the system performance and fabrication techniques need to be improved to reduce the warping of the devices.

5.4.2 128 × 128 LCOS-based compact optical correlator module

This section describes an optical correlator which uses the 128 × 128 LCOS EASLMs, described in Section 5.2.4, as the input and filter planes (Serati, Ewing, Serati, Johnson, & Simon, 1993). These SLMs feature higher SBWP, smaller pixel pitch (30 μm), and faster frame rate (~ 5 kHz) as compared to the 64 × 64 LCOS EASLMs discussed in the previous section. Experimental evaluation of this optical processor demonstrates an improved image correlation performance over the prototype 64 × 64 LCOS-based optical correlator.

Compact optical correlator design

The optical correlator described here was developed by Boulder Non-linear Systems and the Martin Marietta Corporation. It comprises a standard PC-AT 386DX computer which contains most of the system electronics and executes the control software, a Sony XC77 electronic camera control unit, a Melles Griot 06 DLD 201 laser driver and the optical correlator head. Within the optical head is a laser diode ($\lambda = 658$ nm) with collimating optics, fully adjustable $4f$ correlator optics, two programmable LCOS EASLMs with interface electronics and a remote CCD camera head. Figure 5.13 is a photograph of the 0.6′ × 1.0′ × 0.4′ optical head.

Software controls most of the correlator functions. The program downloads data to the SLM frame memory which can store 512 frames per SLM. Each frame is randomly accessed under the control of the correlator software. During program execution, the system software receives instructions from a sequence instruction file. This instruction file sets the width and amplitude of the laser pulse, determines which of the 512 frames is written to the SLM and defines which correlation outputs are captured and stored. Programmable information from the sequence instruction file along with other selectable parameters controls the system's processing cycle. A typical processing cycle for 500 Hz operation is shown in Figure 5.14. The load time indicated is

Figure 5.13 The compact LCOS EASLM correlator.

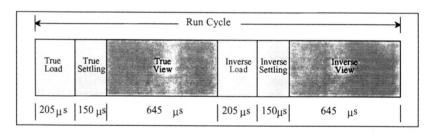

Figure 5.14 Processing cycle for a 500 Hz correlator frame rate.

the time required to write a single frame completely, which is 205 μs. The liquid crystal settling time is set globally and usually equals the FLC's response time, $\tau_{lc} = 150$ μs.

As the correlator processes information at 500 Hz, the CCD camera, which transfers data at standard video rates, cannot continuously capture the optically processed data. However, the CCD camera has a fast responding asynchronous shutter which is electronically triggered by the system program. By electronically shuttering the camera, a single 500 Hz process cycle is captured and stored once per second. The camera is triggered by the system computer at the start of a frame. Its electronic shutter remains open through one complete processing cycle. After the camera has captured the desired frame, the data are transferred to a frame grabber. The frame grabber can locally store four frames at the standard 30 Hz frame rate, but writing the frames to disk eventually slows the system to one frame per second. The asynchronous shutter allows the correlator system to capture single frames and analyze the system's high speed performance.

The optical configuration shown in Figure 5.15 is a fully programmable VanderLugt correlator. The LCOS EASLMs are located at the focal planes of the Fourier transform lens. The image data displayed on the reference SLM are illuminated by the collimated ($D = 7.5$ mm) laser beam ($\lambda = 657$ nm) and are Fourier transformed by the lens. The filter SLM, illuminated by the transform from the input SLM, is encoded with a filter function that multiplies the spectral components of the image data. The filter is generated and implemented as a BPOF. The modulated light from the filter SLM is transformed by the second combination lens resulting in the convolution of the image data with the filter pattern. An image pattern that matches the filter will produce a collimated wavefront that is focused to a bright spot in the correlation plane where the CCD camera is located. The positions of the bright spots coincide with the locations of the matching patterns in the input image. Phase distortion (spreading of the correlation peak) is

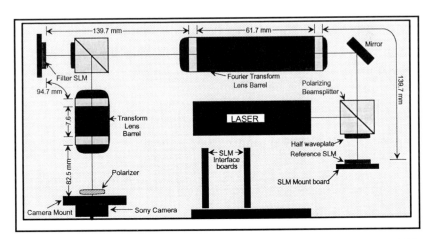

Figure 5.15 Schematic diagram of the 4f optical correlator using two LCOS SLMs.

minimized since both the filter SLM and the correlation plane are at the focal planes of the inverse transform lens. The half-wave plates and PBSs control the polarization state of transmitted light in the system allowing for optimal phase modulation at the EASLMs. The polarizer P2 at the front of the camera is used to increase the extinction ratio of the correlation-detection leg since broadband beam cubes have low extinction ratios when used in reflection-mode.

Experimental results

Correlation performance data were obtained from the optical correlator using a set of test imagery. The objects were located on- and off-axis and disguised by varying degrees of clutter. The images contained binarized trucks, planes and the letters X and O (see Figure 5.16).

With the system operating at 500 Hz, the correlator's output was captured and analyzed. Figure 5.17 shows three-dimensional plots of correlation output using a binarized on-axis truck pattern (Figure 5.16(c)). These three-dimensional plots give a visual comparison of the correlator's 500 Hz performance (Figure 5.17(a)) with respect to the performance of an ideal system (Figure 5.17(b)). The peak-to-secondary represented by Figure 5.17(a) is 7.65 dB whereas the theoretical peak-to-secondary shown in Figure 5.17(b) is 16.37 dB. Tables 5.2 and 5.3 summarize some results from the correlator's 500 Hz performance tests. Table 5.2 shows the system's detection sensitivity in comparison to the theoretical peak-to-rms noise (PRMS) and peak-to-secondary (P/S) for the objects and filters listed in the table.

Table 5.3 presents results from tracking a plane across a cluttered background (refer to Figure 5.16(b)). The location and orientation of the plane changes as it 'flies' from left center, through the center and out the top center of the background frame. In the first two frames of the sequence, the plane was partially obscured by a strip which masks off reflective areas outside of the array. Therefore, these frames are not included in the Table. This test demonstrates that the SLMs have sufficient uniformity to track accurately off-axis objects. Correlation peak accuracy was determined to be within a pixel across the input.

Although the optical correlators described in this section succeeded in performing pattern recognition, there are several improvements in the SLM technology that will enhance the system performance, as discussed in the next section.

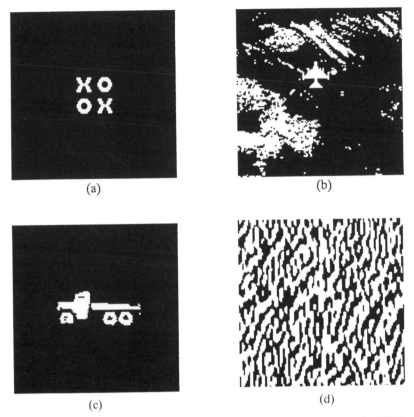

Figure 5.16 Examples of test imagery for evaluating the correlator's 500 Hz operation: (a) small Xs and Os, (b) plane against cluttered background, (c) binarized on-axis truck, and (d) BPOF of the truck.

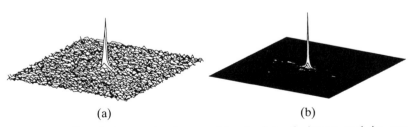

Figure 5.17 Correlation peaks from a binarized truck image and its corresponding BPOF: (a) three-dimensional plot of CCD output from 500 Hz performance data, and (d) three-dimensional plot of theoretical performance.

Table 5.2 *500 Hz sensitivity data from performance testing using Sony camera.*

		Experimental		Theoretical	
Input	Filter	PRMS (dB)	P/S (dB)	PRMS (dB)	P/S (dB)
Truck (5c)	Truck filter (5d)	10.67	7.65	27.13	16.37
Small Xs and Os (5a)	Little O filter	11.46	6.69	20.24	7.83
Small Xs and Os (5a)	Little X filter	10.96	4.95	20.33	8.18
Plane (5b)	Vertical plane filter	8.13	5.12	17.08	7.52

Table 5.3 *500 Hz off-axis object tracking accuracy using Sony camera.*

Input plane sequence	Filter BPOF of plane	Experimental peak location	Theoretical peak location
3	Horizontal plane	90,90	90,90
4	Horizontal plane	80,81	80,80
5	Horizontal plane	70,70	70,70
6	Horizontal plane	Missed frame	60,60
7	Vertical plane	60,50	60,50
8	Vertical plane	55,40	55,40
9	Vertical plane	49,30	50,30

5.5 Future research in high speed optical correlators

5.5.1 EASLM directions and technology improvements

LCOS smart pixel devices will continue to mature with the development of its two constituent technologies, silicon VLSI and liquid crystal displays. Smaller devices and/or arrays with larger numbers of pixels are expected as the minimum integrated circuit dimensions continue to shrink. As minimum dimensions decrease so will the maximum available voltage to the liquid crystal, hence decreasing liquid crystal switching speeds. Furthermore, yield issues will become more dominant. In the following we briefly describe improvements in these two technologies, and their implications for LCOS Smart SLMs.

Improvements in silicon

Scaling Smaller feature sizes in the integrated circuit process will result in increasing the pixel fill factor and also the number of pixels that can be fabricated on a die whose size is limited by the processing aperture of photolithography stepper systems. Since the optical efficiency of an optical correlator scales as the fill factor to the fourth power, one can expect a major increase in system performance by going to the smaller geometry process. These improvements are attained at the expense of lowering the value of the maximum voltages that can be switched by transistors in the smaller line process, resulting in a slower liquid crystal response time.

Planarization For coherent optical processing applications of the LCOS technology, it is desirable for assembled SLMs to be at least $\lambda/4$ if not $\lambda/8$ in flatness. A smooth substrate surface also allows for better alignment and thickness control of the liquid crystal layer. As the layers in the integrated circuit (IC) process are deposited, they destroy the flatness of the chip by introducing a topography that can vary across the die by 3 μm or more. Furthermore, the chip is subjected to thermal cycling conditions during fabrication and can warp. Planarization of the die has been achieved by oxide deposition using an electron cyclotron resonance (ECR) system to produce a thick layer of silicon dioxide, which can be polished flat. Contacts to the metal modulating pads are made by etching through the oxide. Another approach would be to spin on inter-metal dielectric (IMD) layers, such as polyimide, to try to fill in the gaps between layers in the chip fabrication. This technique has been successfully used with the DMD technology.

Metalization The quality of the metal layers in the IC process could be improved to reduce their grain size, and prevent hillocks from forming on the metal mirrors that act not only to provide the appropriate voltage to the liquid crystal pixel, but also as a reflective substrate for the light reading the pattern of the EASLM. The addition of a small amount of copper to the usual silicon/aluminum mirror reduces the metal layer grain size but makes the composite difficult to etch. The hillocks can be removed by either processing the chip at a lower temperature or adding an oxidation layer prior to sintering.

Silicide For large EASLM arrays the resistance of the polysilicon bus line can cause a significant increase in the electronic loading time, τ_{elec}, of the device. This can be alleviated by the use of lower resistance silicided polysilicon which has a resistance that is ~ 6 times lower than that of ordinary polysilicon (Sze, 1988).

Passivation A passivation layer is usually deposited as the last layer of the IC process to protect the circuitry from being scratched or damaged in handling the post-processed wafers. For the LCOS technology, it is necessary to define an etch-back in this layer such that the liquid crystal can make contact with the metal modulating pads. This opening in the overglass reduces the pixel fill factor, due to the design rules requiring a minimum of a few micrometers overlap between the modulating pad metal etch-stop and the passivation layer. In practice, the opening in the passivation layer is not completely etched as designated, resulting in a fill factor on the previously discussed 128×128 EASLM of 30% as compared to the 55% specified in the chip layout. Elimination of the passivation layer maximizes the use of the mirror although it leaves the chip more vulnerable to handling damage and general atmospheric degradation.

Packaging Reduction of backplane warpage, as highlighted in Section 5.3.4, is important for improving EASLM performance. Low stress epoxies (Jared, 1993) and solder bump bonding (Lin *et al.*, 1993) are both under investigation as possible solutions to warpage problems.

Yield Yield is one of the key issues for the LCOS EASLM technology. Memory chip designers build redundant lines of memory cells along the edge of their arrays so that, in the case of a failure, data can be rerouted to a good line. The use of redundancy (Bisotto, Micollet & Poujois, 1985) and conservative layouts are currently the best approaches to maximizing yield for LCOS devices.

Simulation and modelling Electrical circuits used in the LCOS devices can be modeled using SPICE or some other widely used circuit simulation program. This modeling allows designers to simulate the performance of their circuitry prior to fabrication of the chip. The equivalent SPICE models for the optical elements including the FLC capacitors, photodiodes, phototransistors, etc., also need to be developed to allow the simulation of the combination of these technologies (Rice & Moddel, 1992). For example, it is desirable to model the influence of exposing the silicon backplane to the read and write optical beams required to operate the SLMs. The incident illumination results in photogenerated carriers that can strongly influence the behavior of the silicon circuitry by altering voltages and currents in various components of the circuits.

Pad reflectivity The reflectivity of a metal pad made of an AlSi mix on an SLM IC was measured to be approximately 40% (McKnight, 1989). This can be improved by better metal processing (O'Hara *et al.*, 1993) or by depositing a high reflectivity dielectric mirror. The thickness of such a layer will have two impacts on the SLM performance, the electric field across the liquid crystal will be reduced, hence the liquid crystal response time will increase, but the pad should be flatter, and more reflective. In addition, the capacitance of the dielectric material may contribute to the electrolytic breakdown of the devices.

High voltage capability One way to increase the response speed of the liquid crystal is to increase the voltage available to the modulating pad. This will be particularly important in smaller geometry VLSI, and can be accomplished by using a BiCMOS process, or lightly doping the transistor drains (LDD). Both allow higher voltages than the 5 V standard CMOS process. Test chips fabricated through the 2 μm MOSIS low noise analog process with LDD have resulted in n-FET voltage thresholds of 30 V and p-FET thresholds of 15 V. These higher voltages would allow $\pm$ 15 V to be applied across the liquid crystal, resulting in 10–30 μs liquid crystal response times (Johnson *et al.*, 1993).

Improvements in liquid crystal technology

The key improvements in liquid crystal technology include liquid crystal alignment techniques and better liquid crystal materials.

Alignment The typical alignments used with liquid crystals include rubbed polyvinyl polymers spun on both sides of the glass plates in simple cells. However, since the silicon backplane replaces one of the glass electrodes, a nonrubbing method is preferred. Obliquely evaporated SiO and MgF are used and are likely to provide an improvement in LCOS alignment (Bawa, Biradar, Saxona, & Chandra, 1990). A good double-sided alignment may allow the contrast ratio and phase modulation uniformity of the LCOS devices to approach those seen in single shutter modulators (> 1000:1).

Material parameters Further improvements in liquid crystal materials include synthesizing chiral smectic liquid crystal mixtures that operate at CMOS compatible voltages with high switching speeds. Inspection of the equation for liquid crystal response time,

$$\tau = \frac{1.8\eta}{P_s E},$$
(5.24)

yields that the switching speed is proportional to the orientational viscosity and inversely proportional to both the spontaneous polarization, P_s, and the applied field E. To achieve fast liquid crystal response times, we need liquid crystals with either low η or high P_s for a given applied field. P_s can be increased up to a limit, but increased values of P_s require an increase in the charge per unit time that must be delivered to a pixel to switch the liquid crystal. Lowering the value of η is more desirable because reduced values of the viscosity parameter improve response time and also reduce power dissipation requirements (Xue *et al.*, 1987).

Material tilt angle The FLC tilt angle, ψ, is especially important in applications where we wish to employ grey level LCOS SLMs. Increasing the tilt angle directly increases device throughput and dynamic range by allowing a greater range of polarization modulation.

Temperature sensitivity FLC SLMs are filled under vacuum at high temperatures ($> 100\,^{\circ}\text{C}$). As the liquid crystal cools, it undergoes phase transitions from nematic, to smectic A (grey level), to smectic C (binary). While operating the SLM, it is necessary to maintain a particular temperature range to prevent transition to a different liquid crystal phase. Certainly, an unwanted phase transition could adversely affect the performance of a system designed for a specific type of liquid crystal modulation. Research in liquid crystals with broad temperature ranges is an ongoing topic in the field of liquid crystal synthesis.

5.5.2 Detector plane considerations

Introduction

At some point the output of an optical processor will be converted to electrical signals for further analysis and/or storage. Detection and post-processing of the optical processor output is usually performed with ordinary video cameras or detector arrays and dedicated hardware and software. As SLM frame rates begin to exceed the rates of even the most advanced detector arrays, new approaches to the detection and processing of optical system output must be designed and implemented. For example, the EASLMs used in the optical correlator systems discussed in the last section operate at frame rates of more than 1 kHz, yet the systems employ CCD cameras that operate at 500 Hz or less.

The photosensitive device at the output of an optical processor may be required to distinguish between intensity levels with great accuracy, or it may only need to respond quickly to the presence or absence of intensity variations. In addition, the information of interest may be the location of some feature in the output or a time history of the intensity variations.

The most commonly used method of optical system post-processing employs a camera or detector that encodes the image intensity information as a video signal. Provided the data rate of the camera is large compared to the rate at which the output changes, the video signal will correctly represent the optical output. Converting the video signal to digital data then allows the acquired information to be processed by dedicated hardware and/or software.

Very high performance image sensors have been developed using charge-coupled devices (CCDs) and photodiode arrays. Both of these consist of arrays of optical detector elements (pixels) with surrounding circuitry for transferring the data from each pixel to an output channel. Typically, such arrays work in two phases. During the first phase, the input optical image is exposed to active pixels producing photogenerated currents or voltages. This period is referred to as the integration time. In the second period, data from all of the pixels are transferred off the array to a computer or dedicated processing circuitry. By interleaving these two phases, devices can perform these two functions in parallel. The rate at which pixel data are moved off the device is v_{pix} and ranges from 5–20 MHz for currently available two-dimensional detector arrays (EG&G Reticon; Dalsa, Inc.). A device with N^2 pixels operates at a frame rate of v_{pix}/N^2. Such detector assays still require that the output be fed into other circuitry for storage and/or post-processing of the desired information. For many applications use of custom microelectronics allows the necessary storage and post-processing computations to be done at the camera data rate. When the system speed of an optical processor exceeds the speed of the output detector array, the detector array becomes the speed bottleneck.

For example, consider the task of locating a small number of bright peaks in a scene, as typically required in optical correlators and ranging systems. Using a moderately sized detector array of 64 × 64 pixels, a 10 MHz data rate allows for a frame rate of 2.5 kHz. Assuming that the post-camera-processing is fast enough the system readout rate will be the camera frame rate. If the modulating devices in the optical system are capable of higher speeds, as is the case with both the deformable mirror and LCOS technologies, the data readout will be the speed bottleneck in the system. Furthermore, if we wish to increase the size of the detector device, the readout speed decreases linearly with the increase in number of pixels. The underlying problem is not the detector technology, it is the method by which the technology is applied.

Traditional readout methods require that the data from each element in the detector array be interrogated before the interesting information can be isolated. For many applications, this type of approach is inefficient. For example, the human visual system does not download intensity information from all of the photoreceptors in the eye before subsequent processing by the brain (Schade, 1956). It appears that pre-processing within the first few layers beyond the photoreceptor array performs a 100:1 compression of the image information sent to the brain.

For the example problem of finding bright peaks in an output image, a detector array using existing technology could be greatly improved if it provided peak locations directly. Reporting ten peak locations at 1 MHz offers two orders-of-magnitude improvement in system speed over using a 10 MHz 64 × 64 pixel detector array that reports intensity information from each pixel. Such an improvement in output processing speed may also be accompanied by a reduced detector cost and easier system integration as multiple electronic boards are replaced with a single application specific integrated circuit detector array.

As we have found with SLM technology, the flexibility of silicon circuit design allows for realization of customized detectors at a low cost. Devices which combine photodetectors and both analog and digital circuitry have been applied to a variety of applications. Biologically motivated tasks, such as early vision processing (Mead & Mahowald, 1988), motion detection (Tanner & Mead, 1984), and auditory processing (Mead, Arreguit, & Lazzaro, 1991) have been implemented using analog silicon circuitry. Devices for determination of position and orientation have been implemented in CMOS for both self-tracking (Bishop & Fuchs, 1984) and external object tracking (Standley & Horn, 1991) applications. Application specific photodetector (ASPD) arrays with on-chip processing offer tremendous design flexibility due to the close integration between optical detectors and electronic circuitry.

An example of ASPD design: the peak location problem

Peak detection as it relates to optical correlators involves two basic problems. The first is to determine if a peak is present, and the second is to determine the location of the peak. In many correlator applications, there can be more than one peak and thus in addition to these two problems some mechanism for dealing with multiple peaks must be included. Two early approaches to the peak location problem have been implemented in CMOS by Chao & Langenbacher (1991a, b) and by Turner, Jared, Sharp & Johnson (1992). Both of these solutions are based on the concept of an array of photodetectors which use inter-pixel competition to sequence through all of the peaks present in a scene, reporting peak locations serially.

The device designed and built by Chao & Langenbacher consists of a 32 × 32 array of pixels located on 200 μm centers. The pixels are globally interconnected, each consisting of a photodetector, which occupies 50% of the pixel area, and digital processing circuitry. The digital circuitry at each pixel causes the pixel to send a disable signal to all other pixels in the array once the photodetector response is above an externally adjustable threshold. Upon reporting the location of the peak the enabled pixel is disabled and the next pixel that is above threshold takes control and makes the new position data available. This process is repeated until no pixels are above threshold. Peak locations are reported in microseconds, although the photodetector response times are significantly slower resulting in device frame times of several milliseconds.

The second CMOS peak-detector implementation is a test chip consisting of a 3 × 3 array of pixels located on 150 μm centers. Again the pixels are globally interconnected, but the global connection acts as an analog competition bus that feeds into a winner-take-all (Lazzaro, Ryckebusch, Mahowald, & Mead, 1988) circuit at each pixel. The winner-take-all function allows the pixel with the strongest response to take control of the output channels while suppressing all of the other pixels. A disabling circuit removes a pixel from the network once it has been a winner and reported the peak location. The device, shown in

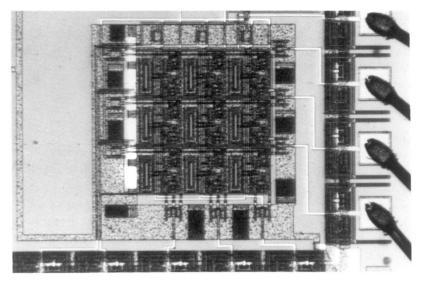

Figure 5.18 3 × 3 peak detector using a winner-take-all circuit. Pixels are located on 150 μm centers. Fabricated in a 2.0 μm CMOS process from MOSIS.

Figure 5.18, can locate a peak in less than 10 μs and can distinguish peak intensity variations of less than 5% over three orders of magnitude in input intensity.

One problem with these two correlation-peak position detectors is that position resolution is much more coarse than the possible peak shifts that exist in the output plane. Multiple peaks in the output of a correlator are the result of multiple input objects. Assuming that the input objects do not overlap significantly, the spacing between peaks must be relatively large compared to the pixel spacing on the SLM used to represent the input. This means that although peaks may shift by small amounts, peak spacings in the output remain relatively large. Some position sensors, such as lateral-effect photodiodes and quadrant cell detectors (Wallmark, 1957; Light, 1986), have the ability to determine the position of a single peak with great accuracy. Using a group or array of ASPDs for peak location, each of which covers the minimum peak spacing at the correlator output, allows accurate peak position information to be determined. Connecting the array of peak locating ASPDs as a competitive sequential peak detector, similar to the two mentioned above, allows the processing efficiency to be retained. This combination of the high resolution single peak detectors and the competitive array devices provides the basic idea for the ASPD that solves the general correlation plane detection problem.

Work is in progress on the design of such devices at the University of Colorado at Boulder (Turner & Johnson, 1994). The correlation peak detector can be designed for fabrication in an ordinary 2.0 μm CMOS process. Each pixel of the device is a relatively large ($\sim$ 400 μm $\times$ 400 μm) position sensor. These sensors can be 2-dimensional lateral-effect or quadrant devices (Wallmark, 1957; Kelly, 1977; Light, 1986; Kawasaki, Goto, Yashiro, & Ozaki, 1990). Output from each of the pixels is fed into an analog position computing circuit and a global competition circuit so that only pixels with peaks will compute the peak locations. The final output of the chip is position data that encode both the local peak position and the coordinates of the pixel that owns the peak. Frame rates for the chip can reasonably be expected to be greater than 10 kHz. Position sensitivity of less than 1 μm in each dimension can easily be realized with detectors that employ quadrant geometry or the lateral photoeffect (Kelly, 1977). Overall, the customized correlation peak detector will provide higher speed and better position resolution than competing detector arrays. These advantages are obtained at lower cost and with a single chip than can be connected directly to a computer or memory device.

5.5.3 Summary

To summarize, we have discussed optical correlators and the use of LCOS EASLMs as well as the application specific photodetectors in the implementation of correlator systems. The influence of various SLM characteristics on correlator performance was investigated, including fill factor, flatness, size, and speed. Two correlator systems employing LCOS EASLMs as input and filter plane modulators were discussed. The first employed 64 $\times$ 64 EASLMs and operated at a maximum frame rate of 3 kHz. The peak-to-secondary ratio for this system was measured to be 2.67 dB. The second system, employing 5 kHz frame rate, 128 $\times$ 128 EASLMs, improved on this performance with peak-to-secondary ratios as high as 7.65 dB.

The LCOS EASLM technology appears to be well matched with the required component and system specifications for optical correlators. These specifications include cost, system size, ease of integration with external electronics, speed, accuracy, and robustness. In each of these categories, LCOS SLMs are seen as comparable or superior to other methods of light modulation. Certainly, improvements in LCOS technology and smart photodetectors are required to exploit fully the advantages offered by optical processing. Many of the tasks and problems associated with this improvement were discussed in Sections 5.3 and 5.5, and specific approaches to solutions were mentioned. SLM improvements including planarization, increased fill factor, and liquid crystal alignment and material enhancements were discussed. Section 5.5.2 focused on a solution to the problem of processing high speed correlator output by employing a combination of global competition and local position sensing photodetectors.

Finally, the existence of proof-of-principle optical correlation systems that are competitive with electronic systems, not only in performance but also in cost, size, and ease of use, is an achievement that strengthens the concept of employing optics and optical devices as a means of improving traditionally all-electronic systems. Optical correlators themselves may only be interesting for specific applications; however, the ability to enhance the performance of current electronic processing systems by exploiting smart pixel technology is interesting on a more general level.

5.6 Acknowledgments

We wish to thank the NSF Engineering Research Center for Optoelectronic Computing Systems and the Martin Marietta Corporation for supporting much of the research in LCOS EASLMs for optical correlation. In addition we appreciate the work of David Jared, Dave Doroski, and Chong Chang Mao on the design and fabrication of the LCOS EASLMs. Technical discussion and assistance from Douglas McKnight, Ian Underwood, and Darren Simon were also enjoyed.

References

Applied Optics special issue on optical inspection, **27** (1988).

T. H. Barnes, K. Matsuda, & N. Ooyama. Reduction of false correlation with binary phase-only filters. *Appl. Opt.*, **27**(18): 3785 (1988).

S. S. Bawa, A. M. Biradar, K. Saxona, & S. Chandra. Novel alignment technique for surface stabilized ferroelectric liquid crystal. *Appl. Phys. Lett.*, **57**(14) (1990).

G. Bishop, & H. Fuchs. The self tracker: a smart optical sensor on silicon. In *Conference on advanced research in VLSI*, pp. 65–73, ed. P. Penfield Jnr, MIT Press, Cambridge MA (1984).

S. Bisotto, D. Micollet, & R. Poujois. Using redundancy when designing active matrix addressed LCDs. In *Proc SID*, **26**: 201–7 (1985).

W. P. Bleha, L. T. Lipton, E. Wiener-Avnear, J. Grinberg, P. G. Reif, D. Casasent, H. B. Brown, & B. V. Markevitch. Application of the liquid crystal light valve to real-time optical data processing. *Opt. Eng.*, **17**(4): 371–84 (1978).

R. M. Boysel. A 128 × 128 frame-addressed deformable mirror spatial light modulator. *Opt. Eng.*, **30**(9): 1422–7 (1991).

R. N. Bracewell. *The Fourier Transform and its Applications*, Chapter 6, p. 98. McGraw-Hill, New York, NY (1965).

D. P. Casasent, & W. A. Rozzi. Computer-generated and phase-only synthetic discriminant function filters. *Appl. Opt.*, **25**(20): 3767–72 (1986).

D. P. Casasent, & S-F. Xia. Phase correction of light modulators. *Opt. Lett.*, **11**(6): 398–400 (1986).

T. H. Chao. Optical correlator using a LCTV CGH filter and a thresholding photodetector array chip. In *OSA Annual Meeting Technical Digest*, volume 17, pp. 150–1, Optical Society of America, Washington, DC (1991a).

T. H. Chao, & H. Langenbacher. Private communication (1991b).

N. A. Clark, & S. T. Lagerwall. Submicrosecond bistable electro-optic switching liquid crystals. *Appl. Phys. Lett.*, **36**: 899–901 (1980).

N. A. Clark, & S. T. Lagerwall. Surface-stabilized ferroelectric liquid crystal electro-optics: new multistate structures and device. *Ferroelectrics*, **59**: 345–87 (1984).

D. R. Collins, J. B. Sampsell, J. M. Florence, P. A. Penz, & M. T. Gately. Spatial Light Modulators and Applications. *OSA Technical Digest Series*, volume 8, p. 102, Optical Society of America, Washington DC (1988).

L. K. Cotter, T. J. Drabik, R. J. Dillon, & M. A. Handschy. Ferroelectric-liquid-crystal/silicon-integrated-circuit spatial light modulator. *Opt. Lett.*, **15**(5): 291–3 (1990).

D. M. Cottrell, J. A. Davis, M. P. Schamschula, & R. A. Lilly. Multiplexing capabilities of the binary phase-only filter. *Appl. Opt.*, **26**: 934 (1987).

W. A. Crossland, & P. J. Ayliffe. A dyed phase change liquid crystal display over a MOSFET switching array. In *Society for Information Display International Symposium Digest of Technical Papers*, volume 12, pp. 112–13, The Society for Information Display, Playa del Rey, CA (1981).

L. J. Cutrona, E. N. Leith, L. J. Porcello, & W. E. Vivian. On the application of coherent optical processing techniques to synthetic aperture radar. *Proc. IEEE*, **54**: 1026 (1966).

J. A. Davis, D. M. Cottrell, G. W. Bach, & R. A. Lilly. Phase-encoded binary filters for optical pattern recognition. *Appl. Opt.*, **28**(2): 258 (1989).

J. A. Davis, M. A. Waring, G. W. Bach, R. A. Lilly, & D. M. Cottrell. Compact optical correlator design. *Appl. Opt.*, **28**(1): 10 (1989).

P. G. DeGennes. *The physics of liquid crystals*. Clarendon, Oxford (1975).

F. M. Dickey, & L. A. Romero. Dual optimality of the phase-only filter. *Opt. Lett.*, **14**(1): 4 (1989).

F. M. Dickey, K. T. Stalker, & J. J. Mason. Bandwidth considerations for binary phase-only filters. *Appl. Opt.*, **27**(18): 3811–18 (1988).

J. D. Downie, B. P. Hine, & M. B. Reid. Effects of correction of magneto-optic spatial light modulator phase errors in an optical correlator. *Appl. Opt.*, **31**(5): 636–43 (1992).

T. J. Drabik, & M. A. Handschy. Silicon VLSI/ferroelectric liquid crystal technology for micropower optoelectronic computing devices. *Appl. Opt.*,

29(35): 5220–23 (1990).

U. Efron, ed. *Spatial Light Modulators and Applications III*, volume 1150. SPIE, Bellingham, WA (1989).

S. Esener. Recent developments on Si/PLZT smart pixels technologies. In *LEOS 1992 Summer Topical Meeting Digest on Smart Pixels*, pp. 85–88, IEEE, New York, NY (1992).

D. L. Flannery, A. M. Biernacki, J. S. Loomis, & S. L. Cartwright. Real-time coherent correlator using binary magnetooptic spatial light modulators at input and Fourier planes. *Appl. Opt.*, **25**(4): 466 (1986).

D. L. Flannery, J. S. Loomis, & M. E. Milkovich. Design elements of binary phase-only correlation. *Appl. Opt.*, **27**(20): 4231–5 (1988).

J. M. Florence. Joint-transform correlator systems using deformable-mirror spatial light modulator. *Opt. Lett.*, **14**(7): 341–3 (1989).

J. M. Florence, & R. O. Gale. Coherent optical correlator using a deformable mirror device spatial light modulator in the Fourier plane. *Appl. Opt.*, **27**(11): 2091–3 (1988).

G. Gheen, & E. Washwell. The effect of filter pixelation on optical correlation. In *Spatial Light Modulators and Applications*, OSA Technical Digest Series, volume 14, pp. 161–164. Optical Society of America, Washington, DC (1990).

P. D. Gianino, & C. L. Woods. Effects of spatial light modulator opaque dead zones on optical correlation. *Appl. Opt.*, **31**(20): 4025 (1992).

J. W. Goodman. *Introduction to Fourier Optics*, pp. 4–29, McGraw-Hill, New York, NY (1968).

J. Grinberg, A. Jacobson, W. Bleha, L. Miller, L. Fraas, D. Boswell, & G. Myer. A new real-time non-coherent to coherent light image converter: the hybrid field effect liquid crystal light valve. *Opt. Eng.*, **14**(3): 217–25 (1975).

S. Heddle, D. G. Vass, & R. M. Sillitto. Reduction of aliasing in correlation using a pixelated spatial light modulator. In *Optical information processing systems and architectures IV*, volume 1772, ed. B. Javidi. SPIE, Bellingham, WA (1992).

L. J. Hornbeck. 128 × 128 deformable mirror device. *IEEE Trans. on Electron Devices*, **ED-30**(5): 539–45 (1983).

L. J. Hornbeck. Deformable-mirror spatial light modulators. In *Spatial Light Modulators and Applications III*, volume 1150, pp. 86–102, ed. U. Efron. SPIE, Bellingham, WA (1989).

J. L. Horner. Is phase correction required in slm-based optical correlators? *Appl. Opt.*, **27**(3): 436–8 (1988).

J. L. Horner. Private communication (1990).

J. L. Horner. Metrics for assessing pattern-recognition performance. *Appl. Opt.*, **31**(2): 165–6 (1992).

J. L. Horner, & P. D. Gianino. Phase-only matched filtering. *Appl. Opt.*, **23**(6): 812–16 (1984).

J. L. Horner, & P. D. Gianino. Applying the phase-only filter concept to the synthetic discriminant function correlation filter. *Appl. Opt.*, **24**: 851 (1985).

J. L. Horner, & P. D. Gianino. Signal-dependent phase distortion in optical correlators. *Appl. Opt.*, **26**(12): 2484–90 (1987).

J. L. Horner, & J. Leger. Pattern recognition with binary phase-only filters. *Appl. Opt.*, **24**(5): 609–11 (1985).

M. Hose. FFT chips for transform-based image processing. *Advanced Imaging*,

56–9, June (1992).

D. A. Jared. *Optically addressed CMOS spatial light modulators*. PhD thesis, University of Colorado at Boulder (1993).

D. A. Jared, & D. J. Ennis. Evaluation of binary-phase-only-filters for distortion-invariant pattern recognition. In *Computer-Generated Holography II*, volume 884, ed. S. H. Lee. SPIE, Bellingham, WA (1988).

D. A. Jared, & K. M. Johnson. Optically addressed thresholding very-large-scale-integration liquid crystal spatial light modulator. *Opt. Lett.*, **16**(12): 967–9 (1991).

D. A. Jared, R. Turner, & K. M. Johnson. Electrically addressed spatial light modulator using a dynamic memory: *Opt. Lett.*, **16**(22): 1785–87 (1991).

B. Javidi, S. F. Odeh, & Y. F. Chen. Rotation and scale sensitivities of the binary phase-only filter. *Opt. Comm.*, **65**: 233 (1988).

B. Javidi, & F. T. S. Yu. Performance of a noisy phase-only matched filter in a broad spectral band optical correlator. *Appl. Opt.*, **25**(8): 1354–8 (1986).

K. M. Johnson, D. J. McKnight, & I. Underwood. Smart spatial light modulators using liquid crystals on silicon. *J. Quant. Elec.*, **29**(2): 699–714 (1993).

K. M. Johnson, D. J. McKnight, C. C. Mao, G. D. Sharp, L. Y. Liu, & A. Sneh. Chiral smectic liquid crystals for high information content displays and SLMs. In *Spatial Light Modulators and Applications OSA Technical Digest Series*, volume 6, pp. 18–21, Optical Society of America, Washington, DC (1993).

R. D. Juday, B. V. K. Vijaya Kumar, & P. Karivaratha Rajan. Optimal real correlation filters. *Appl. Opt.*, **30**(5): 520–2 (1991).

R. R. Kallman. Optimal low noise phase-only and binary-phase-only optical correlation filters for threshold detectors. *Appl. Opt.*, **25**: 4216 (1986).

R. R. Kallman. Direct construction of phase-only filters. *Appl. Opt.*, **26**: 5200 (1987).

A. Kawasaki, M. Goto, H. Yashiro, & H. Ozaki. An array-type PSD (position-sensitive detector) for light pattern measurement. *Sensors and Actuators A: Physical*, **22**: 529–33 (1990).

B. O. Kelly. Position sensing with lateral effect photodiodes. In *Effective Utilization of Optics in Quality Assurance*, volume 129, pp. 59–62, ed. J. Cruz. SPIE, Bellingham, WA (1977).

H. M. Kim, J. W. Jeong, M. H. Kang, & S. I. Jeong. Phase correction of a spatial light modulator displaying a binary phase-only filter. *Appl. Opt.*, **27**: 4167–8 (1988).

C. J. G. Kirkby, M. J. Goodwin, & A. D. Parsons. PLZT/silicon hybridised spatial light modulator array-design, fabrication and characterisation. *Int. J. of Optoelectronics*, **5**(2): 169–78 (1990).

A. V. Krishnamoorthy, G. Yayla, & S. C. Esener. A scalable optoelectronic neural system using free-space optical interconnects. *IEEE Trans. on Neural Networks*, **3**: 404–13 (1992).

B. V. K. Vijaya Kumar, & Z. Bahri. Phase-only filters with improved signal to noise ratio. *Appl. Opt.*, **28**(2): 250–7 (1989).

B. V. K. Vijaya Kumar, & L. Hassebrook. Performance measures for correlation filters. *Appl. Opt.*, **29**(20): 2997–3006 (1990).

A. I. Lakatos, & R. F. Bergen. TV projection display using and amorphous-Se-type ruticon light valve. *IEEE Trans. on Electron Devices*, **ED-24**(7): 930–4 (1977).

J. Lazzaro, S. Ryckebusch, M. A. Mahowald, & C. A. Mead. Winner-take-all networks of $O(n)$ complexity. *Neural Information Processing Systems*, pp. 703–11, ed. D. S. Touretzky, Morgan Kaufmann, San Mateo, CA (1988).

L. Leclerc, Y. Sheng, & H. H. Arsenault. Rotation invariant phase-only and binary phase-only correlation. *Appl. Opt.*, **28**(6): 1251–6 (1989).

S. H. Lee, S. C. Esener, M. A. Title, & T. J. Drabik. Two-dimensional silicon/PLZT spatial light modulators: design considerations and technology. *Opt. Eng.*, **25**(2): 250–60 (1986).

W. Light. Optical position sensing using silicon photodetectors. *Lasers and Applications*, **5**, no. 4, 75–9 (1986).

W. Lin, Y. C. Lee, & K. M. Johnson. Study of soldering for VLSI/FLC spatial light modulators, pages 491–7 in *1993 Electronic Components and Technology Conference (ECTC)*, IEEE, New York, NY (1993).

H. K. Liu, J. A. Davis, & R. A. Lilly. Optical-data-processing properties of a liquid-crystal television spatial light modulator. *Opt. Lett.*, **10**: 635 (1985).

A. VanderLugt. Signal detection by complex spatial filtering. *Trans. IEEE*, **IT-10**, 139–45 (1964).

D. J. McKnight. *An electronically addressed spatial light modulator*. PhD thesis, University of Edinburgh (1989).

D. J. McKnight, D. G. Vass, & R. M. Sillitto. Development of a spatial light modulator: randomly addressed liquid-crystal-over-nMOS array. *Appl. Opt.*, **28**(22): 4757–62 (1989).

D. J. McKnight, K. M. Johnson, & R. A. Serati. 256×256 liquid-crystal-on-silicon spatial light modulator. *Appl. Opt.*, **33**(14): 2775–84 (1994).

C. A. Mead, X. Arreguit, & J. Lazzaro. Analog VLSI model of binaural hearing. *IEEE Trans. on Neural Networks*, **2**(2): 230–6 (1991).

C. A. Mead, & M. A. Mahowald. A silicon model of early visual processing. *Neural Networks*, **1**: 91–7 (1988).

R. B. Meyer, L. Liebert, L. Strzelecki, & P. Keller. Ferroelectric liquid crystals. *J. Phys. Lett.*, **36**: L69–L71 (1975).

F. Mok, J. Diep, H. K Liu, & D. Psaltis. Real-time computer-generated hologram by means of liquid-crystal television spatial light modulator. *Opt. Lett.*, **11**(11): 748–50 (1986).

W. E. Nelson, & L. J. Hornbeck. Micromechanical spatial light modulator for electrophotographic printers. In *Proceedings of SPSE Fourth International Congress on Advances in Non-Impact Printing Technologies*, p. 427, ed. A. Jaffe, Society for Imaging Science and Technology (formerly the Society of Photogramic Scientists and Engineers), Springfield, VA (1988).

A. O'Hara, J. R. Hannah, I. Underwood, D. G. Vass, & R. J. Holwill. Mirror quality and efficiency improvements of reflective spatial light modulators by the use of dielectric coatings and chemical-mechanical polishing. *Appl. Opt.*, **32**(28): 5549–56 (1993).

D. R. Pape, & L. J. Hornbeck. Characteristics of the deformable mirror device for optical information processing. *Opt. Eng.*, **22**(6): 675–81 (1983).

D. Psaltis, E. G. Paek, & S. S. Venkatesh. Optical image correlation with a binary spatial light modulator. *Opt. Eng.*, **23**(6): 698–704 (1984).

R. A. Rice, & G. Moddel. A circuit model for the optical and electric response of an FLC to an arbitrary driving voltage. In *Society for Information Display International Symposium Digest of Technical Papers*, volume 23, pp. 224–7, The Society for Information Display, Playa del Rey, CA (1992).

W. E. Ross, D. Psaltis, & R. H. Anderson. Two-dimensional magneto-optic spatial light modulator for signal processing. *Opt. Eng.*, **22**(4): 485–90 (1983).

J. B. Sampsell. An overview of the digital micromirror device (DMD) and its application to projection displays. In *Society for Information Display International Symposium Digest of Technical Papers*, volume 24, pp. 1012–15, The Society for Information Display, Playa del Rey, CA (1993).

O. H. Schade, Sr. Optical and photoelectric analog of the eye. *J. Opt. Soc. Am.*, **46**(9): 721–39 (1956).

S. A. Serati, T. P. Ewing, R. A. Serati, K. M. Johnson, & D. M. Simon. Programmable 128 × 128 FLC SLM compact correlator. In *International Symposium on Optics and Photonics in Aerospace and Remote Sensing, Optical Pattern Recognition IV*, volume 1959, ed. D. P. Casasent, SPIE, Bellingham, WA (1993).

S. E. Shields, B. G. Fletcher, & W. P. Bleha. A 240 × 320 element MOS addressed high resolution LCD. In *Society for Information Display International Symposium Digest of Technical Papers*, volume 14, pp. 178–9, The Society for Information Display, Playa del Rey, CA (1985).

D. Standley, & B. Horn. Analog CMOS IC for object position and orientation. In *Visual Information Processing: From Neurons to Chips*, volume 1473, pp. 194–201, ed. B. P. Mathur and C. Koch. SPIE, Bellingham, WA (1991).

S. M. Sze. *VLSI Technology*. John Wiley and Sons, New York, NY (1988).

J. E. Tanner, & C. Mead. A correlating optical motion detector. In *Conference on advanced research in VLSI*, pp. 57–64, ed. P. Penfold Jnr. MIT Press, Cambridge, MA (1984).

J. T. Tippett *et al.* Optical and electro-optical information processing. In *Symposium on optical and electro-optical information processing technology*. MIT Press, Cambridge, MA (1965).

R. Turner, D. A. Jared, G. D. Sharp, & K. M. Johnson. Analysis and implementation of an optical processor employing cascaded VLSI/FLC spatial light modulators and smart pixel detector arrays. In *LEOS 1992 Summer Topical Meeting Digest on Smart Pixels*, pp. 13–14, IEEE, New York, NY (1992).

R. M. Turner, & K. M. Johnson. CMOS photodetectors for correlation peak location. *IEEE Phot. Tech. Lett.*, **6**(4): 552–4 (1994).

R. M. Turner, D. A. Jared, G. D. Sharp, & K. M. Johnson. Optical correlator using VLSI/FLC electrically addressed spatial light modulators. *Appl. Opt.*, **32**: 3094–101 (1993).

I. Underwood, D. G. Vass, & R. M. Sillitto. Evaluation of an nMOS VLSI array for an adaptive liquid-crystal spatial light modulator. *IEEE Proc*, **133**(1): 77–82 (1986).

J. T. Wallmark. A new semiconductor photocell using lateral photoeffect. *Proc. IRE*, **45**: 474–83, (1957).

G. S. Weaver, & J. W. Goodman. A technique for optically convolving two functions. *Appl. Opt.*, **5**(7): 1248–9 (1966).

J. Xue, M. A. Handschy, & N. A. Clark. Electro-optic response during switching of a ferroelectric liquid crystal cell with uniform director orientation. *Ferroelectrics*, **73**: 305–14 (1987).

6

Optical and mechanical issues in free-space digital optical logic systems

F. B. McCORMICK and F. A. P. TOOLEY

AT & T Bell Laboratories, Department of Physics, Heriot-Watt University

6.1 Introduction

This chapter discusses the issues involved in the design of a free-space digital optical logic system. The first section is an introduction to this field with an overview of the justification of pursuing this exciting and challenging course. The second section is a discussion of the basic characteristics of the devices that must be used in a system of this type with specific examples of the behavior of three of the most popular devices. Section 6.3 is a discussion of the optical and mechanical constraints involved in the design process. The fourth section is a design example using logic gates made from symmetric self electro-optic effect devices (S-SEEDs). The final section summarizes the likely future directions.

6.1.1 Background of digital optics

Two methods exist for communicating high bandwidth information: light and electricity. It is clear that the optimum choice for long distances (>1 km) is light; hence the adoption of fiber for telephony. It is equally clear that over short distances (<1 mm), the optimum choice is electricity hence its use for gate-to-gate interconnection in a chip. It has been shown that as the distance between the transmitter and receiver increases, the energy required by an electrical connection increases more rapidly than if optics is used (Miller, 1989). It is difficult to justify the use of optics over short distances and conversely, electrical connection is not ideal except over the shortest distances. The ultimate optimum choice for chip-to-chip and board-to-board interconnection is unclear.

220

There is, thus, interest in 'enhancing' the interconnection of electronic logic using optics. There may be some, as yet unidentified, combination of circumstances (number of connections, bandwidth, interconnection distance, fan-out, noise and ground loop immunity, reliability, enhanced functionality, cost, weight and size) where optical interconnection is the most appropriate choice. At the most basic level, individual light emitting diodes (LEDs) and lasers coupled into a fiber connected to a detector are already being used in high performance computers to provide some of the connections between frames. Small one-dimensional (1-D) arrays of fiber ribbons are also available. There is research into producing an optical backplane which provides the connections between shelves using guided waves in, for example, polyimide.

Free-space optical interconnections represent one possible role that light may play in achieving massive connection capacities (more than 1000 high speed connections) as in Figure 6.1. Free-space may ultimately provide larger interconnection density than conventional electronic techniques. Free-space potentially enables many global or non-local interconnections to be made easily. These are of a form which might be called 'connector-less' interconnections since with the optical pin-outs, no physical connection need be made to the chip. The existence of a technology with such a large number of large bandwidth ports enables the system architect to have the freedom to consider novel partitioning. Thus the optical pin-out may have a large impact on architecture rather than just being another option which is judged in competition with established techniques.

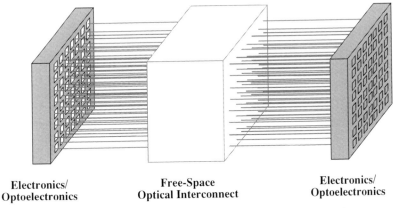

| Electronics/ | Free-Space | Electronics/ |
| Optoelectronics | Optical Interconnect | Optoelectronics |

Figure 6.1 Massively parallel free-space optical interconnection.

The potential benefits of free-space optical interconnects include:

(i) *Higher interconnection densities* with lower power dissipation (and thus fewer heat removal problems).

(ii) *Less signal distortion and dispersion* partially due to easier impedance matching.

(iii) *Less crosstalk* and sensitivity to electromagnetic interference (EMI).

(iv) *Less skew* of system clock and data signals due to the constant delay achievable with free-space optical interconnects (FSOI).

(v) *Higher utilization of chip areas* for processing rather than communication. Because FSOIs take advantage of the third dimension above the chip surface to implement the interconnects, the area currently used to route lines may be used to increase the gate density and chip functionality.

(vi) *Higher reliability* may be a result of lowered thermal loads and noncontact connections.

(vii) *Lower costs* compared to equivalent systems using purely electronic interconnects, due to simplified packaging requirements for high performance systems.

The tradeoffs and costs associated with implementing FSOIs will form the bulk of the rest of this chapter. The primary limitations are:

(i) *Alignment and packaging* of optical transmitters, optical channel elements, and optical receivers.

(ii) *Optoelectronic transmitter and receiver* cost, reliability, and fabrication issues.

(iii) *Loss* of optical power along the interconnection path will be unavoidable. In complex prototype systems, this loss has been on the order of 10–20 dB. Advanced fabrication optical system techniques and monolithic system packaging may improve these figures substantially.

(iv) *Delay* of the signals, although constant across all connections in FSOIs, is unlikely to be less than in electrical interconnections. The propagation delay of FSOIs is slightly less than for electrical connections, but the physical path lengths may be longer due to the need to incorporate lenses, mirrors, beamsplitters, etc.

As interconnection speed and densities increase within an electronic system, at some point the tradeoffs between optical and electrical interconnection begin to favor FSOIs. Most electrical interconnect problems worsen with increasing connection length, while FSOIs are, in most ways, length independent. Hence, an optimal combination appears to be to perform local connections electrically and global connections optically. Furthermore, many of the costs associated with FSOI are weak functions of the number of channels. Indeed, for the optoelectronic transmitters and receivers, monolithic integration of 2-D optoelectronic device arrays should significantly lower costs, while improving device reliability. Thus the cost per channel will be lowest in systems utilizing large numbers of high density FSOIs. Since this is a problem area for electrical interconnects, it seems a reasonable application area in which to develop this new technology.

Several applications require many high density, high speed interconnections. Multistage interconnection networks (MINs) and self-routing switching fabrics used in telecommunications switching can lower blocking probabilities and increase network reliability by increasing the connectivity between stages (Cloonan, Richards, McCormick, & Lentine, 1991). At the high data rates demanded by information-age services, this creates a critical need for high density, high speed interconnects. Certain computing architectures, especially those performing pattern recognition and image processing tasks, are essentially I/O bound. They need many connections between sensors, processors, and memories at moderate-to-high data rates (Kiamilev, Esener, Ozguz, & Lee, 1990). Since FSOIs are unlikely to lower signal propagation delays, architectures without stringent latency requirements will benefit most from FSOIs. Synchronous machines with pipelined architectures and highly asynchronous machines are both potential candidates.

6.2 Overview of optical logic device characteristics

6.2.1 General characteristics

This section outlines the characteristics of devices which may be considered for use in a free-space digital optical system. The use of free space allows the use of 2-D arrays of devices. Consequently, this discussion is limited to devices which can operate as large arrays (>1000 channels). Devices which are designed for use with waveguides, such as lithium niobate directional couplers, are therefore excluded. Devices which have been considered to be potentially useful include: (i) all-optical devices which operate as optically bistable switches, (ii) hybrid devices such as self electro-optic effect devices (SEEDs), and (iii) emitting devices such as laser diodes and pnpn optical thyristors.

The simplest optical system is a point-to-point link with no fan-out, fan-in or interconnect. An emitter or modulator acts as one optical port which must be connected to another port which acts as a digital detector, i.e., the detector and any electronic circuit to which it is connected must determine whether a logic one or logic zero has been sent. A more interesting use of optics may take advantage of the ease with which nonlocal interconnection and fanning can be achieved, and require logical decisions to be taken by the optical ports.

Hence, free-space digital optics may need optical devices to play the same role as a transistor in conventional electronics. In addition to being the optical port, i.e., a modulator or emitter and a detector, they may also be required to act as a thresholding device. Such devices are called optically bistable (OB). These are optical elements which, over some range of light input powers, have two possible output states (Figure 6.2). The range of powers over which they are bistable corresponds to a region of hysteresis and is bounded by two discontinuities at which switching between the two states can occur. Such switching can be induced by optically biasing the element close to one of these discontinuities and then making an incremental change in the total light input. This power increment can derive from an independent (signal) input. It is possible to obtain a change in output larger than the signal input and hence achieve digital gain.

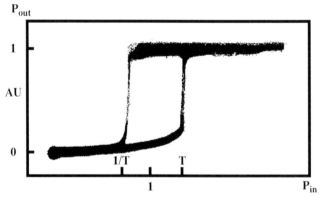

Figure 6.2 Bistable power transfer characteristic of a nonlinear Fabry–Perot resonator.

Since the first demonstration of devices of this sort in the late 1970s there have been a large number of experimental OB systems reported (Walker, 1991; Wherrett, Walker & Tooley, 1988). Optically bistable devices can be subdivided into four main classes (Figure 6.3). The first division is between active and passive. Active devices are those in which the total light output can be more than the total light input (i.e., those having an internal gain stage or light source). Passive devices have no light generating or amplifying components and consequently only produce signal gain by transferring a signal from one low power beam to another of higher power. In these the power source is in the form of an optical input. A common conceptual difficulty is understanding how 'passive' devices, i.e., devices which are not themselves a source of optical energy, can provide any kind of optical gain. 'Active' devices, such as laser diode amplifiers or a device such as a phototransistor driving an LED, can clearly provide real optical gain since they convert some other source of power (e.g., electricity) into light. The key is that 'passive' devices, just like the transistor in electronics, can induce changes in power in one beam with smaller changes in power in another. Such passive devices require some external optical power source to generate the beam that will be modulated, but it may be possible to run many such passive devices from a single light source.

Both active and passive OB devices can be further divided into

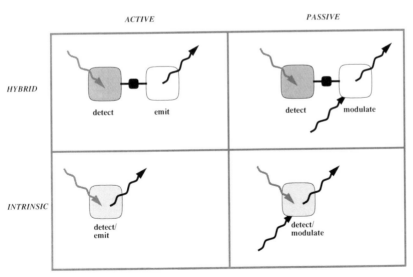

Figure 6.3 Device categorization: active vs passive, and hybrid vs intrinsic.

hybrid and intrinsic. A hybrid device is one in which the positive feedback required for optical bistability is provided electrically. This may involve some external circuitry or be achieved by the internal electrical characteristics of the device. The switching speeds of hybrid devices are usually ultimately limited by the time constants of the electrical feedback. Intrinsic OB devices are constructed from optically nonlinear media and rely on either direct optical feedback or the nature of the nonlinearity itself. In this case switching speeds are determined by medium time constants. Hybrid is a term which applies to those devices involving dissipative electronic transport. Hybrid devices may appear to be inefficient since energy is converted from optical to electrical and back to optical limiting the performance of such a device. In fact, the efficiency of hybrid devices depends on the degree to which they are integrated; there is nothing fundamentally inefficient about photodetectors, laser diodes or some modulators. The inefficiency comes about in practice from the energy required for electrical communication over macroscopic distances from one device to another. The integration of every component is a technologically demanding task.

The selection of the most appropriate device for any application is dependent on many system issues and cannot be decoupled from those issues. However, the device characteristics which are important in this choice are now outlined and explained.

Device size

Detector size The only advantage of a large detector is that packaging will be made cheaper and easier since the alignment of the system is less sensitive to aberrations, misalignment, and instability.

The advantages of a small detector are:

(i) Less crosstalk will occur: the origin of this includes scattered light, unused orders from diffractive components, back reflections and polarization leakage.

(ii) Higher density of optical connections will be possible due to the small pitch associated with small device size.

(iii) There will be more room for electrical components such as logic and traces.

(iv) The detector area is proportional to capacitance, so small detectors will lead to lower switching energy and therefore higher speed for a given signal power.

(v) Another advantage of a small detector is that there should be low intrinsic noise (Tooley & McCormick, 1991).

Modulator/emitter size The advantage of a large size for a modulator is the same as that for the detector – namely increased tolerance to misalignment and use of larger spot sizes. The advantage of using a large emitter is that the output power will be increased.

The advantages of using a small modulator/emitter are:

(i) Less capacitance which may be the limiting factor in the system speed.
(ii) Fewer defects which could cause a device to be nonoperational will be encountered.
(iii) In the case of microlasers, the current threshold will be reduced and the laser can be single tranverse mode which will improve the coupling efficiency.

Pitch

It is assumed that at least 1000 pixels are required. The advantages of using a small pitch are:

(i) The chip yield will increase since the size of chip required will be small and therefore less defects will be encountered.
(ii) If a conventional 'bulk' lens is used to illuminate the devices (i.e., a compound lens mounted in a barrel), then the smaller the pitch, the better. This is because the field over which such lenses can attain adequate performance (adequate is defined in Section 6.3) is limited to around a few millimeters. Consequently, the number of channels that can be operated is inversely proportional to the square of the pitch.
(iii) The latency of electrical signal propagation may be dependent on pitch.
(iv) If phase gratings are used to provide array generation, the fabrication requirements of minimum feature size are relaxed if the pitch is small.

The advantages of a large pitch are:

(i) Thermal considerations, since the pitch must be kept large to avoid thermal crosstalk. In addition, the total thermal load on the device must be kept small and large pitch may help in this respect.

(ii) In addition, optical crosstalk will be lessened.

(iii) More room will be available for electrical traces and components.

Switching energy, power, and speed

These three issues are all related. In general, it is possible to trade speed for power whilst keeping the energy constant. It is often not meaningful therefore simply to quote the lowest power or highest switching speed alone. What must be considered is the total energy input, the product of switch time and switch power.

$$\tau_{sw} = E_{sw}/P_{sig} \tag{6.1}$$

and

$$P_{sig} = P_{laser}\eta_{sys}/N \tag{6.2}$$

so,

$$\tau_{sw} = E_{sw}N/P_{laser}\eta_{sys}, \tag{6.3}$$

where τ_{sw} is the switching time, E_{sw} is the switching energy, P_{sig} is the absorbed signal power, P_{laser} is the output power of the power supply laser, η_{sys} is the optical system power efficiency, and N is the number of beams into which the power supply laser is divided. For example, with $N = 1000$, $\eta_{sys} = 10\%$ and $P_{laser} = 100$ mW (typical of a single mode laser diode) then $P_{sig} = 10$ μW and a switching energy of below 100 fJ is required if 100 MHz operation is needed. Given a limited amount of laser power (P_{laser}), and our desire to operate a large number of logic gates (N), to increase system speed we must increase the system's light efficiency and/or lower the device switching energy. These two solutions generally conflict, since the main way to lower the switching energy is to decrease the device size, and this can decrease the amount of light coupled into the device. For example, the optical switching energy of a S-SEED (see Section 6.2.2) is (Lentine et al., 1989a,b):

$$E_{sw} = CV/S, \tag{6.4}$$

where C is the device capacitance, V is the bias voltage, and S is the device responsivity. Now, $C \approx 0.1$ fF/μm^2 in these devices, thus halving the device diameter will lower the switching energy by a factor of 4. In practice, this approach in S-SEEDs is limited by two effects (see Section 6.2.2). As the device size decreases, photoelectron recombination and parasitic capacitances begin to dominate the switching energy, so that further decreases in size have less effect. Additionally, the optical power loss due to imperfect coupling, or clipping of the signal spots by the devices will act to *increase* the switching time. This loss of efficiency is highly related to other aspects of the optical and mechanical system design and will be discussed in the next section.

Often potentially misleading results are quoted for the switch time of optical devices. These are times that are impossible to take advantage of in engineered systems. The switch time of an S-SEED could be as short as tens of picoseconds (Boyd *et al.*, 1990) which is limited by the diffusion time of photogenerated carriers across the junction. However, such a switch time is only possible if the device is illuminated by a high power, short duration pulse. If the input to a device originates as the output of a similar device, such performance is not possible. One reason for this is that the nonlinearity which is the origin of the bistable behavior saturates at high power. So although a high power, short duration pulse can be used to measure the switch time, such a pulse cannot, in general, be generated as the output of a previous S-SEED device. The practical switch time is hardly influenced by the diffusion time. The effect is 'second order' – the saturation intensity is linked to the intrinsic time, which therefore allows higher powers to be used as the read beam (Boyd *et al.*, 1990).

Another circumstance in which confusion can occur is the practice of quoting switch 'on' times for a device but not switch 'off' times. Often the switch off time is limited by carrier recombination which can take up to microseconds to occur. What is important is the maximum data rate that could be used. A complication of this issue is due to critical slowing down. This is an effect that limits the maximum switch time of intrinsic bistable devices (Janossy *et al.*, 1986). When the switch power is insufficient to overdrive the device, the switch power is typically 2–10 times as long as the energy/time/power tradeoff would suggest.

Data representation

There are several ways in which digital data can be represented optically. The most straightforward is the power level or energy level in the beam. This is used in most systems (e.g., nonlinear etalons and microlasers) since it is the most direct, and it is the only property of light that can be measured by detectors. This is sometimes referred to as single-rail logic to differentiate it from dual-rail logic in which the ratio of two power levels is used to represent a bit. Such a differential data representation is used for S-SEEDs using two spatially separated beams and therefore detectors. Another data representation technique is time (the presence or absence of light during a time slot). This technique was used in experiments with soliton switching in birefringent fibers.

Other ways in which digital information can be encoded onto light include polarization, phase, wavelength, spectral content, size, position, shape, and direction. This list is probably not complete but as far as we are concerned here, all of these data representation methods reduce to the ability to measure power and compare it to some threshold. In particular, several devices work by polarization encoding (e.g., all liquid crystal devices) but the 'decision' is made after the polarization has been analyzed.

Temperature

It is undesirable to operate a device at anything other than room temperature. Just as important as the operating temperature is the range of temperatures over which the device will operate. It may be the case that there is unequal thermal loading of the device at the center and edges of the array. A laser bar is a good example of this. It is found with such devices that the center laser operates at a higher temperature than those lasers on the outside of the bar since the latter only have heat sources on one side of them. The problem may be much worse with 2-D arrays. The temperature coefficient of wavelength is a particular concern with sources, especially if diffractive optical elements are used in the optical system. The temperature of devices when operated individually may be different to that when an array is operated, so the characterization of the response at different temperatures is important. Some materials such as polymers and liquid crystals may degrade at high temperatures. Condensation is a problem with all materials at a temperature below the dew point, since the water may form on the window obscuring the light or may damage the material (GaAs SEEDs are particularly sensitive to water damage). It is probably necessary that optical devices be operated at a constant room temperature.

Output contrast

The contrast is the ratio of the output powers in the two states. It can range from *circa* 2:1 for a S-SEED to >1000:1 for emitters which radiate no light when switched off. High contrast is desirable but not necessary. In general, for reasons of tolerance to nonuniformity, the contrast should be higher than the number of signals that are fanned-in. It is unlikely that devices with a contrast of less than 2:1 would be practical.

Noise margin

Another important device characteristic is its optical 'noise margin' of tolerance to signal power variations. One source of signal power variations is laser power fluctuations. The use of differential signal encoding and a single laser power supply can make the system insensitive to these temporal optical power changes, however, *spatial* power fluctuations may still exist. This spatial signal power nonuniformity will be due to space-variant optical system loss. Thus the device characteristics mainly influence the optical/mechanical design via their requirements on the system loss mechanisms. The system design issues are largely a result of the balancing of various loss tradeoffs.

Throughput

Throughput is important as it affects the data rate as shown by Eq. (6.3). The loss due to the device can be quite high for absorption modulators. For example, a S-SEED used at the optimum wavelength will modulate its reflectivity from around 40% to 10%, a >3 dB loss. Other modulator devices operate by changing phase and can, in principle, be less lossy. Emitter devices on the other hand obviously have no loss. Coupling onto and off the emitters or modulators may introduce additional loss (see Section 6.3.1).

Operating wavelength

Three issues are involved in the discussion of wavelength: (i) optics, (ii) device availability, and (iii) source availability.

Optical considerations favor operation in the infrared. The reason for this is that the performance of diffractive optics is enhanced at longer wavelengths. State-of-the-art is a minimum feature size of ~ 1 μm with a placement accuracy of > 0.1 μm. This is adequate for high efficiency multilevel gratings for wavelengths of this feature size (i.e., 850 nm and longer). In the mid-infrared (~ 10000 nm), exotic glasses such as ZnSe must be used for which microlithographic techniques are not well developed.

Devices can, in principle, be fabricated that operate at any wavelength. A disadvantage of operation at long wavelengths (3000–10000 nm) is that cooling would be required since 'kT' at room temperature is comparable to the photon energy. Devices that operate in the ultraviolet and visible may take advantage of the potentially superior packing density. In addition, the detectors may be thinner due to the small absorption length. However, there may be fabrication problems as semiconductor growth for these large bandgaps is not well developed.

Source availability is a further impediment to the use of ultraviolet or visible devices since no efficient reliable laser exists, except low power laser diodes which operate at ~ 680 nm. It is likely that room temperature II–VI laser diodes and doubled GaAs diodes will become available, but cost and reliability due to the high photon energy causing damage will be an impediment to progress. CO_2 lasers and solid-state lasers are well established and potentially cost-effective sources for infrared wavelengths. However, they are not readily modulated at high frequency and low cost. Telecommunication systems use 1300 and 1550 nm diode lasers but these are expensive and low power. High power GaAlAs laser diodes operating around 800–850 nm are reliable, cheap, mature, and easily modulated.

Beam combination and fan-in

The device array characteristics determine much of the optical channel geometry. Four basic optical system geometries are illustrated in Figure 6.4. The simplest FSOI geometry is shown in Figure 6.4(a), where each device array has detectors on one side and emitters on the

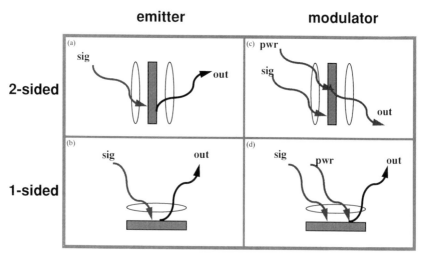

Figure 6.4 Device configurations: emitter vs modulator, and two-sided vs one-sided access.

other side. In this case separate optical systems are dedicated to imaging onto the detectors and collecting the ouput signals from the emitters. A more complicated case is that of Figure 6.4(b), where only one side of the device array is accessible, and the detectors and emitters both reside on that side. Both the input and output beam arrays must now share the same optical system. A similar situation is shown in Figure 6.4(c), but in this case the device array uses modulators rather than emitters. For modulators another optical 'access port' must be supplied and the optical system of one side must be shared even with 'two-sided' device arrays. The last case is that of reflection-mode modulator (Figure 6.4(d)), where the lone optical system is shared between both inputs and the output. The 'one-sided' cases may be the most common in the near term, since they generally are the easiest to fabricate, and provide access to the substrate for mounting and heat sinking. These four cases were described for a fan-in of 1. Larger fan-ins are possible. Multiple signal inputs might each come from a different device array or the interconnection operation may generate more than one signal beam array (fan-out). Thus, in all but the simplest case, we need to provide multiple 'access ports'.

Two-port access is the minimum multiplexing or sharing of the optical system that may be required. For example, we may want to image two inputs or one input and one output onto one side of a device array. The output beam might be the output from an emitter, or the reflection of a power supply input beam from a modulator. Combination or separation of two beam arrays can be done in several ways:

(i)   Amplitude division. A simple 50:50 beamsplitter located at a system pupil will combine two beam arrays, but wastes half of the incident power.

(ii) Polarization multiplexing. If a polarizing beamsplitter is used, most of the loss may be avoided, provided that the two beam arrays are linearly orthogonally polarized.

(iii) Wavelength multiplexing. Similarly, if the beam arrays are of two different wavelengths, a dichroic beamsplitter can be used with low loss. However, to achieve low loss over a range of propagation angles usually requires some minimum wavelength separation (≈ 20–$40\,$nm for $7°$ range of propagation angles in each array).

(iv) Propagation angle (in pupil or image plane). In the pupil, an angularly selective element like a volume hologram can, in principle, combine beam arrays propagating at different angles.

In practice, fabrication problems, crosstalk, and reduced efficiency limit this approach. At the image plane, the beams in the
two arrays may be focused at different angles, by *spatially*
separating the beams in the pupil plane. This is sometimes
referred to as 'pupil division' (Walker, 1986a,b). Because the
beam arrays do not use the full aperture of the optical system,
the resulting spot sizes are increased. In the absence of aberration effects, if circular beams of half the pupil diameter are used,
then the resultant spot diameters will be twice the size of spots
formed by the full aperture. Another potential problem is the
lack of telecentricity in this technique. This causes the spot
positions to move if they are defocused slightly, and introduces
small amounts of time skew between channels.

(v) Spatial position (in pupil or image plane). Similarly, the image
plane may be divided to interlace arrays of focused spots. This
may be accomplished using patterned mirrors. One set of inputs
is reflected off the mirror array, while the other counter input set
is transmitted through the mirror array. Thus combined, they
both are imaged into the device array by the same lens, but at
adjacent spatial positions.

Note that these schemes simply share the optical system, they do not
necessarily combine multiple input beams into the same mode (Goodman, 1985). This is not possible without loss. The inputs in these
schemes are not combined onto the same mode, in general they differ
in the spatial location of the focused spots (e.g., different detectors),
propagation angle (e.g., pupil division), wavelength or polarization.

Interface with electronics

The interface with conventional optics is one of the most important
issues that has to be considered when comparing the relative merits of
different devices. The all-optical systems are very poor in this respect.
An interface such as a spatial light modulator is required to inject the
information into the system and to control the function of each gate in
each device plane. Ideally, a technology would be integratable with
silicon. Because of its indirect bandgap, silicon is unsuitable as an
emitter or as a low voltage modulator. There is much current interest
in four approaches:

(i) Monolithic cointegration of III–V semiconductor optical
devices with silicon electronics (Dobbelaere, Huang, Unlu &

Morkoc, 1988; Goossen *et al.*, 1989; Barnes *et al.*, 1989). Monolithic cointegration of optical devices with silicon CMOS will efficiently leverage mature CMOS technology and allow very high levels of circuit complexity. Near-infrared transmitters enable the use of silicon detectors. GaAs multiple quantum well modulators with long lifetimes ($> 10\,000$ hours) have been grown on silicon (Goossen *et al.*, 1989). Forward-biased devices such as LEDs and lasers have also been grown on silicon, but have suffered very short lifetimes. A likely failure mechanism is nonradiative recombination leading to fatal propagation of dark-line defects. For high yields of any cointegrated structures, progress to achieve CMOS-compatible low temperature growth and processing techniques is also necessary.

(ii) Monolithic cointegration of electro-optic or polarization modulators with silicon electronics. Significant progress has been made in the cointegration of PLZT ($PbLaZrTiO_3$) and ferroelectric liquid crystal (FLC) modulators with silicon electronics (McKnight, Vass & Sillito, 1989; Drabik & Handschy, 1990; Lee, Esener & Drabik, 1986). Both PLZT and FLC are materials which can be used to change the polarization or phase of light through the application of an electric field generated by a voltage on the output of an electronic gate. Both of these modulators have the very attractive characteristics of high transmission and contrast. For PLZT modulators, the relatively high switching energies required and the fairly complex fabrication limit the application of these devices. FLC/silicon modulators have been integrated in large arrays, but are presently only capable of low speed (< 100 kHz) operation (Lee *et al.*, 1986).

(iii) Monolithic cointegration of III–V semiconductor optical devices with III–V semiconductor electronics (Nichols *et al.*, 1988; Miller *et al.*, 1989; Olbright *et al.*, 1991; Chan *et al.*, 1991). Large scale integration of optical and electronic devices in GaAs has yet to be demonstrated, but integration of single detectors, transistors, and modulators/emitters has been achieved. This approach is attractive since it may enable much higher reliability optical and electronic devices, lower costs via monolithic integration, and potentially very high speed systems.

(iv) Hybridization of optical devices to silicon (Goodwin *et al.*, 1991). Hybridization of optical devices with silicon electronics requires very minimal changes of CMOS processing techniques and allows independent optimization of the optical and electrical device properties. Controlled collapse chip connection (C4), or 'flip-chip' bonding techniques using solder-bumps are commonly used to achieve the electrical connections between the electronic and optical devices. Solder-bump bonding has been used to align three planes (silicon electronics, InP optoelectronics, and micro-optics) to less than 2 μm tolerances. Vertical alignment to sub-micrometer precision has been reproducibly demonstrated (Parker, 1991). Epitaxial liftoff techniques (Yablonovich, Gmitter, Harbison, & Bhat, 1987, 1989) may also be applied to this hybridization task.

6.2.2 Specific examples (Tooley & Lentine, 1992)

Nonlinear etalons

The first OB device to be described will be the nonlinear Fabry–Perot etalon (NLFP) (Wherrett *et al.*, 1988; Janossy *et al.*, 1986; Tooley & Lentine, 1992; Felber & Marburger, 1976; Wherrett, 1986; Karpushko & Sinitsyn, 1982). A Fabry–Perot etalon is composed of two highly reflective mirrors that are separated by a cavity as shown in Figure 6.5. Of the impinging light on the etalon, usually only a small portion will enter the cavity. The rest will be reflected. The light that enters the cavity encounters the second reflective surface. At this interface, a portion of the light escapes. The remaining light is now propagating in the opposite direction until it is reflected by the initial reflective surface where the majority of the light will reverse direction within the cavity and another small part escapes from the cavity. If the round trip

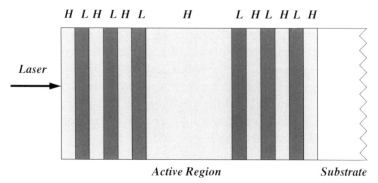

Figure 6.5 Nonlinear Fabry–Perot resonator schematic.

phase delay of the cavity is equal to an integer multiple of the wavelength of the light in the medium then constructive interference will occur between the light that has been propagating in the cavity and any new light that is entering the cavity. When this resonance condition occurs, large intensities of light can be built up within the cavity.

Suppose that the cavity is filled with a material which has an index of refraction that varies with intensity. This can be represented by $n = n_0 + n_2 I$ where n_0 is the linear index of refraction, n_2 is the nonlinear index of refraction and I is the intensity of light within the cavity. Since the peak resonance of the etalon is a function of the index of refraction of the cavity, at low cavity intensities the resonant peak could be at the same wavelength as that if a linear medium were present in the etalon. As the intensity increases the resonant peak shifts (to longer wavelengths, if n_2 is positive). With an incident beam of wavelength longer than the cavity resonance the device will initially be reflective. As the intensity is increased the resonant peaks of the cavity will shift increasing the transmission of the device. The effect of a nonlinear index of refraction on the reflectivity of an etalon as a function of input intensity is shown in Figure 6.2. As the intensity of the incident light increases, so does the intercavity light intensity which shifts the resonance peak. This shift in the resonant peak increases the transmission which, in turn, reduces the reflectivity. This reduction will continue with increasing intensity until a minimum value is reached at which power it will start increasing.

The characteristic curve can be used to approximate an all-optical NOR gate. When no inputs are present the output is a 'one' while the presence of any input will force the output to a 'zero'. To implement a NOR gate function using the characteristic curve shown in Figure 6.2 requires a third input, the bias beam. This energy source biases the etalon at a point on its operating curve such that any input will exceed the nonlinear portion of the curve. This technique is called critical biasing.

The first system in which passive intrinsic optical bistability in a semiconductor was observed was in the form of an interference filter constructed by depositing a series of thin layers of transparent material of various refractive indices on a transparent substrate (Karpushko & Sinitsyn, 1982). The two outer stacks have the property of high reflectivity at one wavelength and thus play the role of mirrors forming a cavity. The nonlinearity is of thermal origin and thus absorption can be introduced by any partially absorbing material forming part of the filter. Although a given filter will only work within a narrow range of wavelengths, different filters of similar construction can be operated at whichever part of the spectrum there is a convenient laser source. The principal disadvantage of thermal nonlinearity based systems is the high switching energy required. By operating with small device volumes low enough switching powers will result ($<$1 mW) to enable simultaneous operation of the order of only 100 pixels (gates) with one laser diode and, simultaneously, switching times of $<$1 ms will be possible.

Bistability in GaAs (and GaAlAs) devices utilizing an electronic nonlinearity was first reported in 1979 (Gibbs *et al.*, 1979). It has been extensively studied since then and continues to evolve into a potentially useful device. The current best reported performance figures are bistability at 4 mW input power with a 25 ns switching time (Olin, 1990). The switching energy is therefore 100 pJ. However, a 16 μm diameter beam was used to obtain the GaAs result so the irradiance is 0.5 mW μm^{-2}. This high irradiance level leads to significant local heating. Typically, about a third of the incident power is absorbed and ultimately contributes to heating. Systems operating at near-infrared wavelengths get so hot that continuous operation of a single channel to minimize thermal effects is usually all that can be achieved. Operation therefore usually requires a low duty factor (100 to 1) which largely obviates the potential advantage of a short switching time. A thermally stable system has recently been obtained by optimizing the cavity design to lower the power required, using a thermally stable sample mount, and only operating one device. Substantial further improvements would have to be made to enable large arrays of closely packed devices to be used in parallel.

NLFPs are normally unpixellated so device size and pitch are freely chosen but $>$ 10 μm diameter beams on $\sim$ 200 μm centers are typical. Single-ended data representation is used and the devices are critically biased. They operate at room temperature with around a degree temperature stability required. The output contrast is around 2 or 3 and the peak reflectivity may be as high as 90%. The operating wavelength is suitable for use with diode lasers. They are commonly used with double-sided or reflection mode.

A wide range of materials has been used to observe intrinsic bistability. The properties of semiconductors have been engineered at an atomic level to try to enhance the nonlinearity. The critical parameter to consider is the product of switch power and switch time. The results of these studies (Walker, 1991; Wherrett *et al.*, 1988; Tooley & Lentine, 1992) show that even using such techniques, all intrinsic OB systems still have a large switching energy. Some OB devices can switch as fast as electronic logic but require so much power that operating more than one channel is not feasible due to thermal constraints and is uncompetitive with the large interconnectivity available with conventional techniques. Other OB systems do have low enough switching powers that operating hundreds of channels simultaneously at rates of around a megahertz is, in principle, possible. However, such operation is also not competitive with electronics which is able to support simultaneous operation of hundreds of pin-outs at hundreds of megahertz.

SEEDs

SEEDs (Lentine *et al.*, 1989a,b; Boyd *et al.*, 1990; Dobbelaere *et al.*, 1988; Goossen *et al.*, 1989; Barnes *et al.*, 1989; Nichols *et al.*, 1988; Miller *et al.*, 1989) rely on changes in the optical absorption that can be induced by changes in an electric field perpendicular to the thin semiconductor layers in quantum well material (Miller, Chemla, & Schmitt-Rink, 1986). A quantum well consists of a material such as GaAs bounded on both sides by a material such as AlGaAs which has a bigger bandgap (the barrier). Many layers may be grown one on another (i.e., barrier, well, barrier, well, etc.) to form multiple quantum wells. If the barriers are thick enough, then the wells act independently and the effect of having multiple wells is to multiply the absorption of a single well by the number of wells. Modern semiconductor growth methods, such as molecular beam epitaxy, enable the well and barrier thicknesses to be precisely controlled to essentially one atomic layer.

In bulk semiconductor materials, there is a smooth absorption spectrum starting at the bandgap energy and rising for increasing photon energy (shorter wavelength). This absorption spectrum near the bandgap changes as we apply an electric field. This effect is known as the Franz–Keldysh effect (Franz, 1958; Keldysh, 1958). In quantum well materials, the electron and hole energies are quantized and large discrete steps in the absorption spectrum are more clearly evident at room temperature. The wavelengths of these absorption steps shift with applied fields. The absorption spectrum (Figure 6.6) exhibits peaks which are called exciton peaks. When a photon is absorbed, a

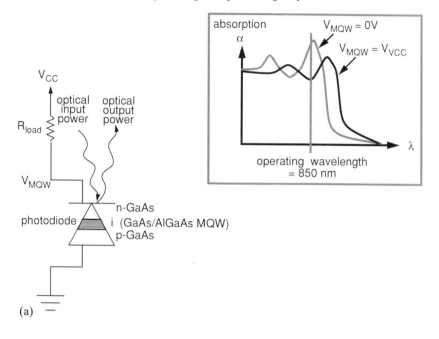

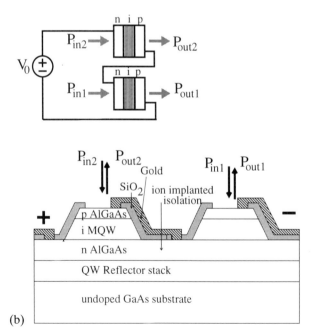

Figure 6.6 SEED: (a) principle; (b) symmetric SEED.

bound electron–hole pair (exciton) is created which does not immediately separate into an electron and a hole but remains together like a hydrogen atom. When an electric field is applied, the electron and hole tend to move towards the opposite sides of the well. However, because of the barriers, the electrons and holes can still orbit around each other and the exciton is preserved. In quantum well material, applying an electric field reduces the photon energy required to create the electron and hole, thus a red shift is seen in the absorption spectrum of quantum well material when a field is applied. This effect is called the quantum confined Stark effect (QCSE) (Miller, Chemla & Schmitt-Rink, 1986; Miller *et al.*, 1984).

The changes in absorption of the GaAs quantum wells are large; 1 μm thick quantum well material can have optical transmission changes greater than a factor of 2 with only 5 V applied. This fact enables arrays of devices to be made with acceptable performance for light propagating perpendicular to the surface of the array. The device structure used most often for SEEDs consists of a reverse biased p-i-n diode with the multiple quantum wells in the intrinsic region. By changing the voltage across the diode directly, the device can be used to modulate electrically an incident beam of light. The basic principle of a SEED is to use the current detected in a photodetector to change this voltage; these devices have optical inputs and optical outputs, even though electrical currents flow within the devices. If the photodetector and modulator are integrated, the SEED can be quite efficient and is one of the lowest energy devices.

The first SEED used a single modulator in series with a resistor, the p-i-n diode modulator also acting as the photodetector (Miller *et al.*, 1988), as shown in Figure 6.6. With no light incident on the photodetector, there is no current, and the power supply voltage appears across the photodiode. As the optical input power is increased, the photocurrent causes the voltage across the photodiode to decrease. If the device is operated at a wavelength where a decrease in voltage causes an increase in absorption, then this increase in absorption causes an increase in photocurrent. This increase in photocurrent causes further reduction in voltage across the photodiode, further increase in absorption, and further increase in photocurrent. This will continue until the quantum efficiency of the photodiode drops off as it approaches forward bias. The net result is that the device switches abruptly from a high to a low voltage state. Since there is increasing absorption with decreasing voltage, this switching from a 'high' to the 'low' voltage state corresponds to the optical output being switched from a 'high' to the 'low' optical state.

Bistable devices such as a NLFP and SEED are effectively two-terminal devices and require precise control of power supply voltages, bias beam powers, and reflections of the output back onto the system. Three terminal devices that avoid critical biasing and provide some input/output isolation are preferable. The S-SEED shown in Figure 6.6 has two p-i-n diodes, each containing quantum wells in the intrinsic region, with one diode behaving as the load for the other (and vice versa). Because the switching of the device depends on the ratio of the two optical inputs, the S-SEED is therefore insensitive to optical power supply fluctuations if the optical beams are derived from the same source. The device has time-sequential gain, in that the state of the device can be set with low power beams and read out with subsequent high power beams. The device also shows good input/output isolation, because the large output does not coincide in time with the application of the input signals. Therefore, the device does not require critical biasing and has the attributes of a three terminal device.

One can extend the S-SEED to get differential devices that can perform any boolean logic function (Lentine *et al*. 1990). The basic concept is that the input light beams are incident on a group of input photodetectors, which may be quantum well p-i-n diodes, that are connected to the center node of a S-SEED. The input photodetectors are electrically connected in a manner that resembles the transistors connections in the CMOS logic family. After the application of the input signals, the voltage on the output S-SEED is either high or low depending on whether or not the logic function is satisfied. These logic SEEDs (L-SEEDs) also retain the desirable qualities of the S-SEEDs; that is, they avoid critical biasing, have time-sequential gain, have good input/output isolation, provide signal retiming, and can be operated over several decades in power level.

A second class of SEEDs has transistors that provide gain between the photodetector and modulator. Two versions of this type of device have been made. One version consists of a heterojunction phototransistor in series with a quantum well modulator (Miller, 1987; Wheatley *et al*., 1987). A second 'amplified' SEED consisted of a quantum well photodiode, a GaAs MESFET (metal–semiconductor field effect transistor) and a quantum well modulator (Miller *et al*., 1989; Woodward *et al*., 1992). In this second example, the quantum well detector and modulator are identical. The top layer of the modulator was made of n-type GaAs and thus the MESFETs were fabricated directly on top of modulators. As optical logic devices, either the phototransistor version of the devices or the MESFET versions are similar. Both devices are single-ended devices that switch when the input(s) exceed a threshold. Differential devices incorporating electronic gain can be made (D'Asaro *et al.*, 1993).

The S-SEED has a fast switching speed of ~33 ps at 20 V increasing to 160 ps at 5 V bias (Boyd *et al.*, 1990). This switching speed is limited by the time it takes for the photogenerated carriers to escape from the quantum wells. Further engineering of the quantum well material may improve this speed further. However, in systems the required optical energy ultimately limits the system speed. The required optical energies of the S-SEEDs that have been made in large arrays are as low as 800 fJ at 5 V bias. While this energy is perhaps too high for ultimate systems with thousands of devices, it is low enough to allow researchers to build prototype systems with low power semiconductor laser diodes. This coupled with the fact that the devices can be made in large arrays and that they are easy to use makes them perhaps the most attractive devices available for use in optical processing experiments.

Electronic gain has been touted as improving the required switching energy of the devices; however, none of the devices that have been described in the literature have accomplished this. For example, in the transistor biased devices described above, the base-collector capacitance needs to be charged directly with the photocurrent, not the amplified photocurrent. This is a direct result of the Miller capacitance that is present in common emitter or common source transistor amplifiers. More sophisticated circuits may be used; for example, the first stage may provide current gain and latter stages voltage gain. However, more complex circuits require more chip area. Also, some of these circuits may reduce the required optical energies, but still require large electrical energies.

6.2.3 Emitters

There are two basic types of active logic device. The first is the light switching 'thyristor-like' device, and the most prevalent of these are the double heterostructure optoelectronic switches (Taylor, Simmons, Cho & Mand, 1989). The second type is the integration of photodiodes, electronics, and lasers, similar to the smart pixels that are the subject of the last section, but using active devices instead of modulators. Only a limited amount of work has been done in this area, and most of it with LEDs as the active device.

The double heterostructure optoelectronic switch (DOES) is a digital active optical logic device with optical output states corresponding to electrical states of high impedance (low optical output) or low impedance (high optical output). The device can be driven from one state to the other either electrically or optically. The optical output can be either a lasing or incoherent output; more prevalent have been the LEDs because they are simpler to make. Since the device is optoelectronic, many functionalities are possible. The DOES consists of an inversion channel heterojunction bipolar transistor (BICFET) integrated vertically with an additional p-n junction. The BICFET is a high performance heterojunction transistor that, unlike heterojunction bipolar transistors, contains no base. The p-channel device contains a thin layer or charge sheet sandwiched between wide gap (AlGaAs) and narrow gap (GaAs) n layers. An inversion channel, a channel in which the dominant carriers are now holes, forms between the two n regions. The purpose of the charge sheet is to act as a source of holes. In the DOES, the BICFET takes the place of a bipolar transistor in a conventional thyristor. Under forward bias in the high impedance state, the collector region is depleted of all carriers, while the collector–substrate p-n junction is forward biased.

Initially, there is the high impedance state, because few carriers flow through the depleted collector and there is a large voltage drop across the collector. Switching begins when excess carriers (holes) are generated in the collector region due to illumination or an increase in voltage. The excess holes generate an even larger increase in electrons flowing in the barrier (emitter) layer because of the gain of the device. Without recombination in the depleted collector, these electrons must cause an increase in electron current of the collector–substrate p-n junction. The hole current in the collector–substrate diode must also increase with the electron current because the external current flowing out of the ohmic contact to the substrate remains small. This hole current now causes a further increase in the barrier hole current, which causes further increase in the barrier electron current. There is positive feedback so long as the collector–substrate hole current density is greater than the holes generated by the input light (or thermally generated holes from increasing the supply voltage). The positive feedback mechanism stops when there are so many carriers in the collector region that they now readily recombine. It is this recombination that is responsible for light emission.

To induce electrical switching in a simpler manner, three terminal devices have been made that use a third contact to the collector of the BICFET (Taylor, Mand, Simmons & Cho, 1987). These devices, known as LEDISTORS or LASISTORS depending on whether the emission is stimulated or spontaneous, use the third terminal voltage to control either the hole concentration in the channel by the injection of electrons into the collector–substrate junction or the charge in the inversion channel directly. Since the charge in the channel controls the current flow in the device, improved performance results if the third terminal is contacted directly with the charge sheet that becomes the inversion channel.

The switching speeds of these devices are impressive, especially considering that all of the devices made to date are quite large (Crawford, Taylor, & Simmons, 1988). For electrical switching, the *RC* time constants determine the speed of the device, provided the third contact (if used) is attached to the channel directly. The optical turn on time is limited by the time it takes the photogenerated carriers to diffuse into the light emitting region. Optical turn off times are also limited by *RC* time constants. For current devices, the *RC* time constants are in the range of 1–10 ns, and optical switch on times were in the tens of nanoseconds. Performance of the devices is expected to improve as the areas are reduced, switching times comparable to the best electronic devices (tens of picoseconds) are possible, although the optical turn on times of at least the surface emitting LED devices will continue to be slower since this time is determined by diffusion effects and not device capacitances and resistances. Lasing devices should offer improved optical turn on times. Switching powers or switching energies have been high because of the large sizes of devices and the fact that all of the surface emitting devices have been LEDs. The required input optical switching energy density can be quite low $(0.02 \text{ fJ}/\mu\text{m}^2)$ if the device without light is biased critically just below threshold. However, the electrical switching energy is much higher.

For all LED based devices, the electrical power dissipation is much too high since currents need to be tens or more milliamperes. Additionally, incoherent light from an LED cannot be effectively collected from small devices or focused onto small devices. Therefore, the LASING DOES or LASISTOR is needed in order to reduce the total power dissipation to acceptable levels. Another approach to active devices might be to combine lasers with electronics and photodiodes, as has been proposed for optical interconnections of electronic circuits. Since the logic function is implemented with electronic circuitry, any functionality can be achieved. Several examples of logic gates have been made using GaAs circuitry and LEDs (Olbright *et al.*, 1991; Chan *et al.*, 1991).

6.3 Optical and mechanical tradeoffs

A basic schematic of an optically interconnected logic system is shown in Figure 6.7(a). In practice, reflection mode devices are often used, leading to a more complex arrangement such as that of Figure 6.7(b). These reflection mode device arrays are generally easier to fabricate and mount. In a digitally regenerating system, we can limit the scope of our design and analysis task to the path from A (generate array) to B (read one device plane) and then to C (write to next device plane). The primary task of the optical and mechanical systems is to image an array of signal beams onto an array of devices, and then image that array of output beams onto the next device plane. For modulator-based systems, we must also generate an array of power supply beams, and focus these beams onto the modulators. A common means of generating these power supply beams is diffractive beamsplitting using

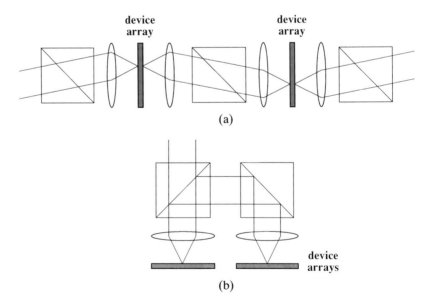

Figure 6.7 Optical processor arrangements: (a) transmissive (b) reflective.

phase gratings (Streibl, 1989; Mair, 1990; Vasara *et al.*, 1992). The reflected, modulated beams form the output signal beam array. These signal beams generally undergo an interconnection operation, which fans out each signal to multiple devices on the next device plane. At this device plane, at least two of the signal beams must be combined with the power supply beams for that device plane, and together they are focused onto the detectors/modulators.

In most systems, the interconnection and beam combination operations add considerable complexity. In general, the optical system must provide high resolution across a wide image field with low image distortion to allow accurate registration of large arrays of small devices. The optical system must also minimize the optical power loss accrued by the required imaging, interconnection, and beam combination operations. The mechanical system is primarily responsible for enabling and maintaining alignment of the lenses, device arrays, and beam arrays. Additionally, it must minimize the total number of components required and provide simple repair and replacement procedures, all in a compact and inexpensive system package. Complicating the optical and mechanical system tasks is the high degree of coupling between them. For example, the imaging quality will be partially determined by the relative alignment of the lenses in the system, while the mechanical alignment precision and stability will be largely determined by the focused spot size, or the optical system resolution. Hopefully, continued investigation and development will allow decoupling of these effects to enable a more modular design approach. However, the highly coupled nature of the different functional pieces of an optical logic system currently requires that they be optimized collectively rather than individually, and in the context of the specific application.

6.3.1 System optical loss mechanisms

Loss of optical power within the system can be attributed to four main mechanisms: Fresnel losses, imaging losses, component losses and alignment losses. Fresnel losses at each surface may be kept to approximately 0.25%/surface by multilayer antireflection (AR) coatings. However, in complex systems having many surfaces, even these small losses may accumulate. For example, a system in which the optical path has 50 AR coated elements will lose almost a quarter of its initial laser power to Fresnel reflections, despite the AR coatings.

Imaging loss mechanisms

The imaging system can have inherent optical power losses due to: clipping at the lenses (often called 'vignetting') or clipping at the device input windows. Clipping at the device windows may be due to lack of sufficient resolution, and/or the presence of excess wavefront aberration. Usually, lenses clip some of the beam power since a perfect Gaussian beam extends indefinitely. The radial intensity distribution, $I(r)$, of a radially symmetric Gaussian beam is:

$$I(r) = I_0 e^{-2(r/w_0)^2}, \qquad (6.5)$$

where r is the radial position, I_0 is the peak intensity and w_0 is the beam radius at which the intensity has fallen to I_0/e^2. The beam diameter is commonly specified as $2w_0$. The power clipped as a function of lens radius, r_{lens}, is:

$$\Delta P/P = e^{-2(r_{\text{im}}/w_0)^2}. \qquad (6.6)$$

Thus we may also specify a 99%-power lens diameter equal to $r_{\text{lens}} = 2(1.52\ w_0)$. Clipping of the focused spot by the device window also introduces a power loss. In the absence of wavefront aberration and significant ($>0.1\%$) clipping by the lens, we can estimate the power lost due to clipping of a round Gaussian spot by a round detector using Eq. (6.6). If the device window is square, the power loss may be calculated by the overlap integral

$$\int_{-r_x}^{r_x} \int_{-r_y}^{r_y} I_0 e^{-2[(r_x+r_y)^2/w_0^2]} \, dx \, dy = I_0 \operatorname{erf}(r_x/w_s) \operatorname{erf}(r_y/w_s), \qquad (6.7)$$

where r_x, r_y are the device window radii, w_s is the spot radius, and $\operatorname{erf}(x)$ is the error function.

A lens with a large aperture focuses a Gaussian beam of radius w_0 to a spot diameter (at the $1/e^2$ intensity points) of:

$$D_{\text{spot}} = \frac{2\lambda}{\pi} \frac{f}{w_0}. \qquad (6.8)$$

In ideal circumstances, 86.5% of the beam's power or energy is contained within the $1/e^2$ diameter of the focused spot. However, for real lenses with finite apertures, the edges of the Gaussian beam are 'clipped', and the resultant truncated Gaussian illumination is focused. When the amount of clipped energy is less than 0.1%, Eq. (6.8) provides accurate results (McCormick, 1993; Wein & Wolfe, 1989; Roberts & Brown, 1987).

When the clipped energy is greater than 0.1%, diffraction effects from the aperture cause a change in the illuminating beam which can be approximated by assuming it is simply a change in waist radius (Belland & Crenn, 1982):

$$w' = w_0 [1 - e^{(r/w_0)^2}], \tag{6.9}$$

where r is the radius (semiaperture) of the lens. This equation assumes that the lens focal length is less than the illuminating beam's Rayleigh range, that the spot size is much smaller than the illuminating beam waist, and that $w' \approx w_0$ which are all valid assumptions for the circumstances in which we are interested. Combining Eqs. (6.8) and (6.9), we have:

$$D_{\text{clipped}} = \frac{4\lambda}{\pi} \frac{f}{2w'} = \frac{4\lambda}{\pi} \left[\frac{r/w_0}{1 - e^{-(r/w_0)^2}} \right] (f/\#),$$

or:

$$D_{\text{clipped}} = \frac{4\lambda}{\pi} \left[\frac{k}{1 - e^{-k^2}} \right] (f/\#), \; k = r/w_0 = \text{clipping ratio}. \tag{6.10}$$

For mildly truncated (between 0.1% and 1% loss) Gaussian illumination, the spot size is dependent not only on the lens $f/\#$ and the beam diameter, but also on the clipping ratio, k, the ratio of lens clear aperture to the beam diameter. An empirical formula for truncated Gaussian illumination has also been developed (Melles Griot Laser Scan Lens Guide, 1987), we have:

$$D_{\text{clipped}} = K\lambda(f/\#) \tag{6.11}$$

where D_{clipped} is again measured at its $1/e^2$ intensity points, and K is a shape factor determined by the clipping ratio, k. An empirical formula for the factor K is

$$K_{-e^2} = 1.6449 + \frac{0.6460}{\left(\dfrac{1}{k} - 0.2816\right)^{1.821}} - \frac{0.5320}{\left(\dfrac{1}{k} - 0.2816\right)^{1.891}}. \tag{6.12}$$

If $k = 1$, then $K_{-e^2} = 1.83$, and for $k = 1.5$, $K_{-e^2} = 2.08$. For clipping ratios greater than about 3.5, the denominators of Eq. (6.12) become complex, and the expression is no longer valid. The 99%-power spot sizes from the analytic and the empirical formulae for 'truncated-beams' (Eqns. (6.10) and (6.12)) match to within about 4% for clipping ratios greater than 1.2.

When focusing energy from a Gaussian beam onto a pixellated detector or modulator, energy may be lost by clipping at the lens and by clipping at the device edges. A Gaussian beam truncated at its $1/e^2$ points will lose 13.5% of its energy. Hence, if we have a clipping ratio of 1 at the lens, and from a spot with a $1/e^2$ diameter equal to the device diameter, we will only deliver 74.8% of the beam's energy. To lessen the clipping at the lens, we must increase the lens diameter, and thus the clipping ratio, k. From Eq. (6.10), however, we see that if the lens $f/\#$ is kept constant, this increase in the clipping ratio will also increase spot diameter, causing more loss at the device. To decrease this loss, we must decrease the lens $f/\#$ and use a 'faster' (and generally more expensive) lens. This tradeoff is illustrated in Figure 6.8, where the beam power delivered to a 5 μm diameter round device window is plotted against the clipping ratio, for several $f/\#$s. Figure 6.9 shows that to couple 99% of a Gaussian beam at 850 nm into a 5 μm device, we need to use about a $f/1.7$ lens at a clipping ratio of about 1.7.

For lower clipping ratios, power is lost due to clipping at the lens; for clipping ratios greater than 1.7, power is lost due to clipping at the detector. Wavefront aberration can increase the spot diameter and cause further loss. Aberration will always be present in any practical system, especially systems such as those which attempt to focus beams to small spots over large image fields. Even the small amount ($\lambda/4$) of aberration present in a 'diffraction limited' Gaussian beam will cause a power loss of around 9%. Aberration can have further deleterious effects on the alignment tolerances (Section 6.1.3).

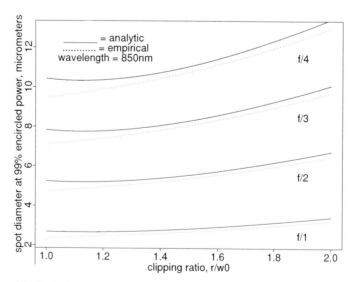

Figure 6.8 Optical system resolution: 99%-power spot size vs clipping ratio for $f/1$ to $f/4$ lenses.

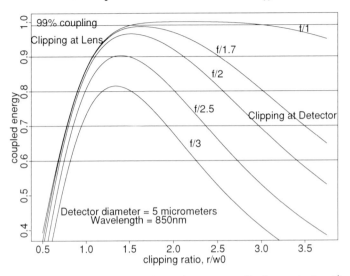

Figure 6.9 Energy coupled into a 5 μm detector vs clipping ratio for f/1 to f/4 lenses.

In light of the difficulty involved in focusing Gaussian beams to small spots in order to couple all of the spot energy into a device of a given size, another design tradeoff is of interest. Since the device switching energy scales approximately linearly with device area (for larger devices), but the spot clipping losses scale at a less than linear rate, an optimal device size should not attempt to collect 'all' of the spot's energy. For large devices, the switching spot is limited by the device area, but as this area decreases, other effects such as parasitic capacitances and recombination are significant. Thus the optimal amount of spot clipping depends mainly upon these 'secondary' effects. Unfortunately, these effects are very device-specific, and have not been widely investigated. Indeed, due to their potential sensitivity to processing variations, the optimal amount of clipping may need to be empirically determined. Two important tradeoffs associated with this clipping are the scattering of the clipped light and, for reflection modulators, the potential degradation of the clipped wavefront.

Component loss mechanisms

Certain components commonly used in these systems can cause additional losses. Binary phase gratings or computer generated holograms commonly direct less than 60–70% of the incident light onto the desired device locations, with the remaining energy being distributed into higher diffraction orders (Streibl, 1989; Mait, 1990; Vasara *et al.*, 1992; Swanson, 1989, 1990). In certain cases, multilevel phase gratings, or volume holographic elements will generally improve the situation, but they can also introduce other limitations. Absorption modulators represent another loss mechanism. For example, simple SEED modulators commonly absorb 50–60% of the incident light, even in the highly reflecting state.

Since most proposed systems utilize polarization effects in some manner, they also suffer from the polarization 'leakage' losses. This misdirection of polarized light can result from several sources. Polarizing beamsplitter (PBS) transmittance and reflectance may vary with angle of incidence to the beamsplitter. These losses may be as high as 10–20% for common catalog precision PBSs. Incorrect fabrication or orientation of polarization retarders within the system will cause further losses. Polarization aberration introduced by many AR coated surfaces is usually small, but in some systems this too may accumulate and become significant (Chipman, 1989).

While not specifically a component loss, the effect of scattered light within an optical logic system can introduce similarly detrimental effects. This is especially true if the scattered light is coherent, in which case as little as 1% scattered optical power can introduce almost 20% variations in the power of the desired signals.

Alignment loss mechanisms (McCormick, 1993)

Power losses may also be introduced by misalignments of the spots on the device windows. To examine the coupling of power from the array of signal beams into the array of detectors (or modulators), we can look at the degrees of freedom of the image planes. Each image plane has six degrees of freedom: translation in x, y, and z (Δx, Δy, Δz), and rotation about the x, y, and z axes (θx, θy, θz). The optical system performance determines the alignment tolerances, and these six degrees of freedom correspond to the optical effects of image shift (Δx, Δy), image rotation (θz), defocus (Δz), and image tilt (θx, θy). Image shift and rotation are basically lateral translation effects, and defocus and image tilt introduce defocus effects.

We can determine the total system tolerance to lateral translation and defocus by examining the power coupling into the detector when the spot is shifted and defocused. For a given misalignment-induced power loss, for example <20%, we can then determine the maximum amounts of shift and defocus. These limiting amounts are then divided among the contributing mechanisms to determine the tolerances on mechanical alignment, wavelength shift, telecentricity, etc. A simple estimate of the tolerances can be made by assuming an unaberrated Gaussian spot of radius w focused onto a rectangular detector of size $D \times B$. The percentage power coupled into the detector for lateral translations of $(\Delta x, \Delta y)$ is the convolution of the spot with the detector, which can be expressed as:

$$\frac{P_{coupled}}{P_{spot}} (\delta x, \delta y) = \left[\text{erf}\left(\frac{D - 2\delta x}{w}\right) = \text{erf}\left(\frac{-D - 2\delta x}{w}\right)\right]$$

$$\left[\text{erf}\left(\frac{B - 2\delta y}{w}\right) - \text{erf}\left(\frac{-B - 2\delta y}{w}\right)\right]. \quad (6.13)$$

A 1-D plot of the power coupling sensitivity for a square 5 μm detector is shown in Figure 6.10(a). A spot with a 99% power diameter of 5 μm nominally couples a little more than 99% of its power into the square 5 μm detector, and this spot can tolerate about a $\pm$ 1.8 μm shift before its coupled power drops below 80%. The defocus limit can be determined by calculating the change in spot size with defocus via:

$$w_z(z) = w_0\sqrt{1 + (z/z_r)^2}, \quad (6.14)$$

and then using w_z in Eq. (6.13) with ($\Delta x = 0$, $\Delta y = 0$). Figure 6.10(b) illustrates the power coupling sensitivity to defocus for a square 5 μm detector. A spot with a 99% power diameter of 5 μm couples 20% less power into a square 5 μm detector when defocused by about $\pm$ 16 μm. Other common defocus tolerance limits come from using the beams Rayleigh range or the Rayleigh quarter-wave criteria $\delta z = 2(f/\#)^2$ (Wellford, 1986).

For our 5 μm spot example, these both indicate about a $\pm$ 10 μm tolerance. These tolerance limits thus result in about 10% loss of power for an aberration-free spot. While this Gaussian beam model is quite useful in a theoretical sense, in practice, the actual tolerances are liable to be worse because of aberrations of the spots. The most accurate means to model these effects is by using a ray tracing program which incorporates diffraction-based point-spread-function (PSF) calculations, encircled energy of the PSF, and convolution of the PSF with a detector. One such program is the optical design ray-tracing package CODE V (*CODE V Introductory User's Guide*, 1989; also Chapter 3). The presence of even a small amount of aberration can significantly affect the coupling efficiencies, as shown in Figure 6.11.

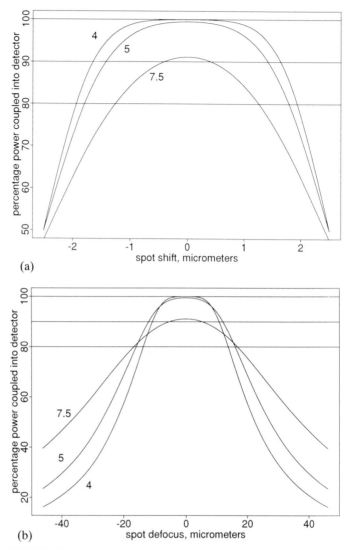

(a)

(b)

Figure 6.10 Ideal power coupling into a 5 μm × 5 μm detector for 4, 5, and 7.5 μm 99%-power diameter spots: (a) with spot shift (b) with defocus.

The simulation which generated the data for these figures used an $f/1.7$ lens and a Gaussian beam clipped (by the lens) at its 99%-power radius. It had 0.56 waves (rms) of spherical aberration and coma and had a Strehl ratio of 0.88, well above 'diffraction limited'. Although the e^{-2} diameter was only 3.9 μm, the 99%-power diameter was over 9 μm due to the small amount of aberration. More importantly, the presence of coma caused an asymmetric distribution of the spot energy, and thus the asymmetric tolerance to lateral shifts. The 80% power loss shift limits for this lens are about +0.9 μm, −2 μm. The defocus tolerance (Figure 6.11(b)) is similarly asymmetric and slightly

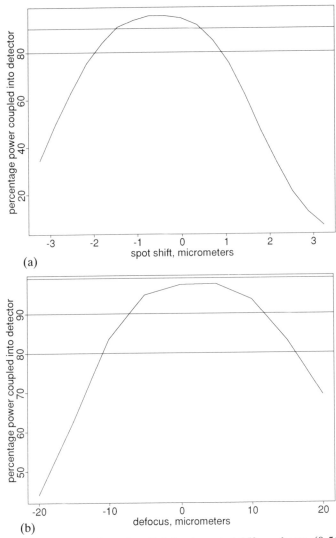

(a)

(b)

Figure 6.11 Power coupling of a slightly aberrated 850 nm beam (0.56 waves rms) into a 5 μm × 5 μm detector: (a) with spot shift (b) with defocus. Spot diameter is 9.1 μm at its 99% encircled power points, and 3.9 μm at its 88% encircled power points. An *f*/1.7 lens having some spherical aberration and coma was simulated (Strehl ratio = 0.883).

worsened ($+16\,\mu$m, $-11\,\mu$m). When aberrations become significant, the alignment tolerancing cannot assume that the effects of defocus and lateral translations are essentially independent. With greater aberration, the spot shapes change as they are defocused, thus the lateral translation tolerances are tightly coupled to the specific focus position.

Defocus and lateral misalignment can be due solely to mechanical movements in the six degrees of freedom listed earlier, but, in practice, optical effects also contribute. Thus the total lateral and defocus tolerances must be divided between all of the potentially contributing mechanisms. This division will be quite specific to the application, optical system design, and optomechanical packaging. Contributors to the overall lateral misalignment budget include:

Telecentricity. In a non-telecentric system, defocus, in addition to increasing spot size, introduces lateral misalignments.

Distortion. For example, when using binary phase gratings (BPGs) for spot generation, the lens distortion must be corrected such that the image height is $f \sin \theta$, due to the diffraction of the BPG.

Focal length. Mismatches in focal length introduce a magnification step between device arrays. To keep this contribution to less than $1\,\mu$m over a 1 mm field, the focal lengths must be matched to 0.1%, or about 20 times better than most catalog precision optics.

Wavelength range. Laser mode-hops and wavelength shifts caused by temperature fluctuations change the beam deflections introduced by diffractive elements. When using BPGs for spot generation, wavelength shifts change the spot spacing. Since this increases additively with the number of spots generated, larger systems are more sensitive to wavelength variations.

Mechanical misalignment in x and y.

Mechanical roll about the z axis.

The rotational alignment tolerance of one array to another also decreases with increasing array size array.

Contributors to the overall defocus budget include:

Image tilt. This can be caused by element tilts and decenters from the lens fabrication and mounting.

Field curvature. This is the curvature of the 'plane of best focus', thus this may be small in systems which artificially flatten their image field by balancing Petzval curvature with astigmatism.

Mechanical tilt about x and y.

Mechanical defocus (z misalignment).

All of the above except x, y, and z misalignments are worst at the edge of the image field, so the worst case tolerances must be specified there. One method of system tolerancing is to let the optical system drive the mechanical tolerances. Optical system simulations can determine nominal values for telecentricity, distortion, image tilt, and field curvature. These, and any other, known values may be subtracted from the overall defocus and lateral misalignment budget, and the remaining tolerance divided equally between the remaining contributors. These tolerance assignments provide a reasonable 'working picture' of the system tolerancing.

6.3.2 System optical nonuniformity mechanisms

Most of the loss mechanisms discussed above may vary in effect across the image field, thus causing signal level nonuniformity. The Fresnel losses will generally contribute the least variation over the low field angles ($<10°$) common to most systems.

Imaging nonuniformity mechanisms

Vignetting at the lens and wavefront aberration will both increase with increasing fields, unless properly compensated for in the system design. Indeed, vignetting is commonly used in the design of microscopic objectives to limit aberration accumulation by clipping portions of the aberrated wavefront, since the reduction in image intensity at the edges of the field is usually not a serious issue in microscopy. This technique is used because certain types of wavefront aberration increase rapidly with field angle (or image height). Coma increases linearly with image height, causing the spot to blur and change shape. Petzval field curvature, and astigmatism increase quadratically in image height, and the combination will also cause blurring and reshaping of the spot's energy. Obviously, the amount of energy coupled into a device window of a fixed size will vary over the field, unless these aberrations are corrected.

Component nonuniformity mechanisms

Uniformity of BPGs used for spot array generation is generally determined by the calculation and fabrication of the phase pattern of the BPG. While this uniformity does not necessarily deteriorate towards the edges of the image, in practice, the individual spot powers vary significantly, especially for large array ($> 32 \times 32$ spots) BPGs with small periods ($< 500\,\mu$m). As discussed earlier, small amounts of coherent scattered light can potentially cause large decreases in the uniformity of the signal levels. The nonuniformity introduced by incoherent scattered light will be directly proportional to the spatial variation of the scattered power.

Variation in the angular performance of the PBS, and variation of the retardance of any waveplates (especially if multiple order retarders are used) may also contribute to the accumulation of nonuniformity. At larger angles of incidence these effects worsen, and other polarization aberration effects such as diattenuation and depolarization become significant (Pezzaniti & Chipman, 1991, 1994).

Alignment nonuniformity mechanisms

Misalignment of the spots onto the devices is often the largest contributor of nonuniformity. Purely lateral (x, y) misalignments will generally affect all signal similarly (in the absence of large aberrations), but the rest of the contributors listed in Section 6.3.1 introduce misalignments which increase with increasing image height, and thus cause approximately radially varying nonuniformity. Imaging distortion increases with the cube of the image height, and thus may be difficult to control. A common technique for limiting distortion is to design imaging systems that are highly symmetric about their aperture stop, so that the distortion effects introduced by the elements closest to the object (e.g., the first device array) are cancelled by opposing effects introduced by the elements next to the image (e.g., the second device array). This is illustrated by the 1:1 telecentric imaging systems of Figure 6.7. Rotational alignment effects are another major source of alignment induced nonuniformity. The long 'lever-arm' of the image radius can require fine rotational adjustment to align precisely the spots to the devices, and the device arrays to one another. For example, for a 2 mm × 2 mm square device array, a ± 1 μm lateral alignment tolerance at the corner of the array requires a rotational alignment precision of 2.4 minutes.

6.4 System design example

As an example of the successful combination of tradeoffs, we will discuss a demonstration of array optical logic using S-SEEDs (McCormick *et al.*, 1991, 1992; Prise *et al.*, 1990). This demonstration represents the 'dumb pixel approach' in that a single device (an S-SEED) acts as detector, logic gate and modulator. The devices had these characteristics: switching energies of 3 pJ; required bias voltage (across all devices in parallel) of 15 V; operational wavelength range of ± 1 nm about 850 nm, or equivalently, a temperature range of ± 5 °C about room temperature; fabricated in square arrays with window separations of 20 μm (since each pair of windows comprises one S-SEED, the device spacing is 20 μm in one direction and 40 μm in the orthogonal direction); window sizes of 5 μm × 10 μm; reflection-mode operation (this facilitates fabrication, improves performance, and allows access to the substrate for optimal heat sinking).

6.4.1 System description

This system cascaded three 16 × 8 arrays of S-SEEDs operating at logic gates. The S-SEED image size at each array was 300 μm × 300 μm, and the S-SEEDs had 5 μm × 10 μm windows on a 20 μm pitch. The crossover interconnection (Jahns & Murdocca, 1988; Tooley, Cloonan & McCormick, 1991) was used in the interconnection unit, and the beam combination unit used space-multiplexing with arrays of small mirrors patterned onto a transparent substrate (Prise *et al.*, 1990; Prise, Streibl & Downs, 1988; McCormick & Prise, 1990).

Optical design

The optical layout of the optical hardware module (OHM) is shown in Figure 6.12(a) and a photograph in Figure 6.12(b). The entire system is 300 mm by 450 mm. A single stripe (index-guided) AlGaAs laser was used which had both a single longitudinal and transverse mode at 850 nm and supplied 3 mW of optical power. The astigmatism of the laser was specified as less than 2 μm. The laser collimating lens had a focal length $f = 5$ mm and a numerical aperture of 0.5. An anamorphic telescope was used to circularize the beams to a typical ellipticity of 1.1:1 and a beam diameter (99% encircled power) of 6 mm. BPGs were used to produce the required 256 beams for each array. The interconnection and beam combination sections use two types of lenses: 42 mm focal length relay lenses and 7.79 mm focal length objective lenses. The relay lenses image the spot arrays onto and off the patterned mirrors and retro-reflector arrays, and the objective lenses image the spot arrays onto and off the S-SEED arrays. The optics in these sections operate over a ± 1.6° field of view.

Along the path from the 16 × 8 BPG to the second S-SEED, six spot

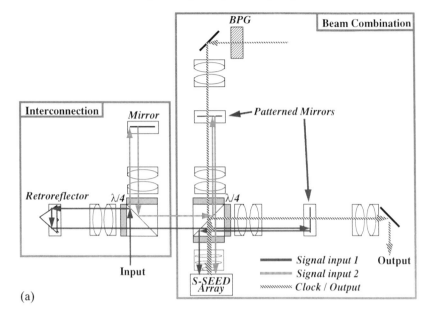

(a)

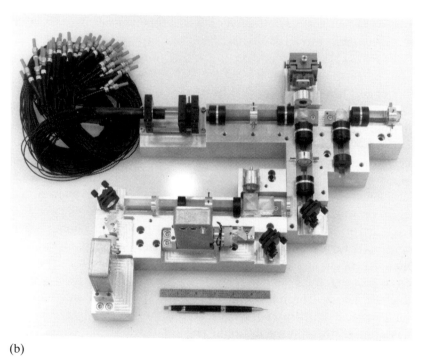

(b)

Figure 6.12 Example system OHM: (a) schematic (b) photograph.

image planes are formed (patterned mirrors, S-SEED, patterned mirrors, retro-reflectors, patterned mirrors, S-SEED). This optical train can be viewed in two ways: either as a series of 1:1 afocal relays, or as a series of telecentric image relays. Each afocal relay transfers the array of collimated beams formed by the 16×8 BPG to the entrance pupil of the next relay. If we examine the image planes instead, we can see that the patterned mirrors, S-SEED arrays, and retro-reflector arrays all lie in conjugate image planes. In these planes, the optical power supply and signals are arrays of focused spots. The imaging between the patterned mirrors and S-SEED arrays has a magnification of 5.4:1 or 1:5.4 (42/7.79). This allows the relay lenses to operate at a higher f-number, facilitating aberration control as well as system alignment and stability.

Along the path from the BPG to the second S-SEED array, the image passes through 11 lenses and a total of 133 surfaces (including beamsplitters, etc.). Eight of the lens passes are through the 42 mm relay lenses, and three of the passes are through the 7.79 mm objective lens. The S-SEED objective lens was an inexpensive catalog laser collimating lens, designed for 820 nm. Two factors in this design help to minimize the total aberration accumulation: system symmetry and the relatively slow speed (high f-number) of the relay lenses. The symmetry of this optical design allowed distortion and coma to be corrected at each relay pair.

The remaining aberration then consisted of spherical aberration, astigmatism, and field curvature contributions. To limit the accumulation of these aberrations, the relay lenses used were of the symmetric (Plossl) eyepiece form. These lenses were constructed from pairs of commercial achromatic doublets. The 42 mm focal length relay lenses were illuminated by beams with a 99% intensity diameter of about 6 mm. Since $1/e^2$ diameter of the Gaussian beam is only about 2/3 of the 99% diameter, the 'effective f-number' (beam diameter/focal length) is greater than for uniform illumination. The relatively high 'effective f-number' operation of the relay lenses resulted in only small aberration contributions from these lenses. Shortening the focal lengths of the relay lenses to less than 42 mm would decrease the system size, but would require a more complex lens form to be designed and constructed. The Plossl relay lenses introduced less than a third of the wavefront aberration of simple achromats, when used in this system. They also provided a means to correct the magnification change that occurs due to the low precision to which the focal lengths of lenses are specified (typically $\pm 1\%$). To ensure registration over the entire array, it was necessary to ensure that every lens was within 0.1% of the correct focal length. This was readily achieved by varying the spacing of the two elements in the eyepiece, via a selection of aluminum spacer rings. The focal length varies approximately as:

$$\text{focal length variation} = \frac{\delta f}{f} \approx \frac{\Delta d}{f_1}\frac{f}{f_2} = \frac{\Delta d}{4f} \text{ when } f_1 = f_2 = 2f,$$

(6.15)

where Δd is the change in the spacing of the achromatic doublets in the Plossl, f_1 and f_2 are the doublet focal lengths, and f is the 'average' effective focal length of the Plossl lens. The spacing, d, is much less than f, (example: $d = 3$ mm, $f = 42$ mm), and relatively large changes in the spacing can finely adjust the lens' focal length.

Mechanical design

The mechanics of this experiment performed three main functions:

(i) Registration of the many lenses to a common optical axis. This was necessary to minimize beam deviation and aberration accumulation.

(ii) Registration of the image planes with respect to each other.

(iii) Alignment of the beam arrays with the common optical axis, and the spot arrays with the image plane elements.

The presence of seven image planes along the optical path from laser i to S-SEED $i + 1$ adds significant complexity to the mechanical system to provide the functionality listed above. To minimize this complexity, the optical components of each OHM were mounted in cells which were placed in V-grooves milled into cast aluminum. Each OHM required only three V-grooves to mount seven lenses, two patterned mirrors and two retro-reflectors. In addition to minimizing the number of mechanical components, the use of a milled plate also ensures that there is a well-defined mechanical axis. The lenses were centered in cells to within 1 arc minute, thus simply placing them into the V-grooves ensured alignment of the optical and mechanical axes.

Registration of the various image planes required positioning in all six kinematic axes. The patterned mirrors and retro-reflectors were mounted in cylindrical $x-y$ positioners that sat in the V-grooves. Focus (z adjustment) and rotation (roll) was obtained by simply sliding and rolling the entire positioner (or lens) along the V-groove. Under a clamping force, focus (even of the S-SEED objective lens) was readily accomplished by sliding the lens cells along the V-groove by hand. Tilt of the patterned mirrors and retro-reflectors was obviated by the large *f*-number (and hence depth of focus) of the spots in those planes. The S-SEEDs were mounted on five-axis positioners located off the plates, as shown in Fig. 6.13.

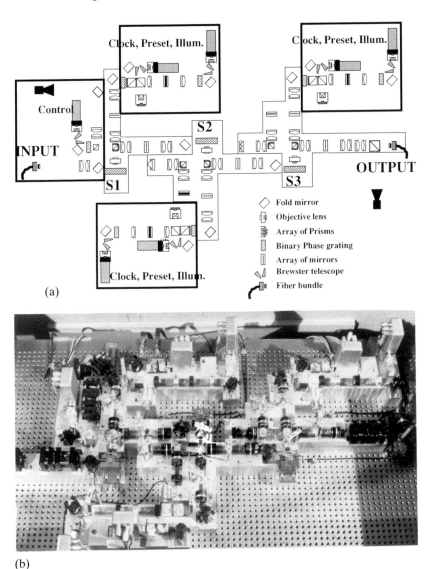

(a)

(b)

Figure 6.13 Complete three-stage system: (a) schematic (b) photograph.

The beams within each OHM and between stages were aligned to the optical axis using fold mirrors. The use of a Twyman–Green interferometer type of arrangement for the interconnection and beam combination subassemblies results in a more compact system and has important alignment advantages. Two beamsplitters separate the paths followed by the two signal beam arrays. The beams will therefore be transmitted through different surfaces and components. However, in both cases, the surfaces not in common are transversed in both directions obviating their effect.

System conclusions

Although the system successfully demonstrated large array optical logic and switching network functionality, its complexity required difficult assembly and alignment and led to relatively large physical size and significant power loss. The main conclusion to be drawn from this experiment is that the practical difficulties resulting from excessive optical and optomechanical complexity can remove much of the theoretical advantages of certain techniques. Much of this complexity was due to the image plane interconnection and beam combination techniques used.

The introduction of image planes to implement these functions has several drawbacks, including increased aberration accumulation, alignment time, overall system size and total component count, as well as decreased system stability, optical power throughput and signal level uniformity. Furthermore, to avoid unwanted image magnification between image planes, the focal lengths of all lenses must be precisely matched. The sources of the loss include imperfect fabrication of the image plane components, aberration and alignment induced coupling losses and polarization leakage losses. Although they are both, in principle, lossless, these techniques result in significant loss (around 50%).

The problems introduced by these image planes (and the associated polarizing beamsplitter assemblies) will generally become more severe as the size of the device arrays increases. Since the tradeoffs between increased aberration, alignment time, system stability etc. are highly coupled, the optimal system design is not obvious. However, it appears that systems operating many thousands of devices per array will have to minimize the total number of imaging steps between device arrays.

The large number of imaging steps make the tasks of aberration control and alignment exceedingly difficult, and result in both loss and signal level nonuniformity. The total system optical power transmission along the per-stage path from laser i to S-SEED $i + 1$ was about 2%. This limited the maximum speed achievable with low-powered diode lasers to 100 kHz. The non-uniformity was about $\pm 20\%$ around the average, as shown in Figure 6.14, which also limits the maximum speed. This is due to the sensitivity of S-SEEDs to signal level nonuniformity when operated as logic gates. Whilst these devices are less sensitive to this than other gates which rely on critical biasing, this tolerance is still a limiting factor.

To increase the feasibility of such systems, two issues must be addressed simultaneously. Firstly, the system complexity must be reduced, lowering the total number of components and especially the number of image planes. This will lower the loss and alignment tasks and cost. Secondly, the device tolerance to signal level nonuniformity must be quantitatively analyzed and highly tolerant operating regimes identified. The identification and analysis of these device-specific 'noise margins' will enable a more practical estimate of the acceptable level of optical/optomechanical complexity.

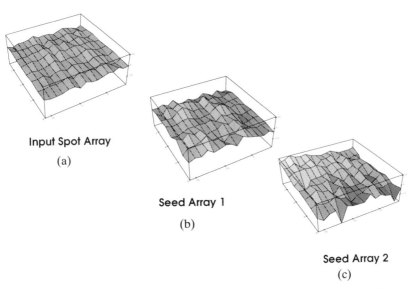

Input Spot Array

(a)

Seed Array 1

(b)

Seed Array 2

(c)

Figure 6.14 Signal power level nonuniformity accumulation: (a) after binary phase grating beam array generation; (b) after first S-SEED array; (c) after second S-SEED array.

6.5 Future directions

Early work in this area has concentrated on the development of optically controlled switching devices (optical transistors). In this chapter we have shown that the optical and optomechanical systems associated with each application can significantly influence the optimal performance characteristics of these devices. The many tradeoffs in device characteristics are highly coupled to each other, as well as to the system design issues. Thus, to be truly feasible, issues such as alignability, stability, cost, reliability must be taken into account when assessing device performance. This is most effectively done via system level modeling and by the implementation of full system demonstration prototypes.

Demonstration systems constructed to this point have interconnected individual logic gates. While enabling the investigation of many critical system issues and allowing a great deal of architectural and optomechanical flexibility, a higher level of pixel integration (or granularity) may provide important advantages. The introduction of appropriate functionality at each pixel (Miller *et al.*, 1989; Feldman, Guest, Drabik & Esener, 1989) may significantly lower the optical system power requirements, the optomechanical complexity and the overall system size. However, decisions concerning the integration of amplification, combinatorial or sequential logic, and memory must be based on the total system tolerancing, since small changes in device performance may have dramatic and widespread system effects.

References

P. Barnes, P. Zouganeli, A. Rivers, M. Whitehead, G. Parry, K. Woodbridge & C. Roberts, 'GaAs/AlGaAs multiple quantum well optical modulator using multilayer reflector stack grown on Si substrate', *Electron. Lett.*, **25**, 995 (1989).

P. Belland & J. P. Crenn, 'Changes in the characteristics of a Gaussian beam weakly diffracted by a circular aperture', *Appl. Opt.*, **21**(3), 522–7 (1982).

G. D. Boyd, A. M. Fox, D. A. B. Miller, L. M. F. Chirovsky, L. A. D'Asaro, J. M. Kuo, R. F. Kopf & A. L. Lentine, '33 ps optical switching of symmetric self electro-optic effect devices', *Applied Physics Letters* **57**(18), 1843–5 (1990).

W. K. Chan, J. P. Harbison, A. C. von Lehmen, L. T. Florez, C. K. Nguyen & S. A. Schwarz, 'Optically controlled surface-emitting lasers', *Appl. Phys. Lett.*, **58**(21), 2342–4 (1991).

R. A. Chipman, 'Polarization analysis of optical systems', *Opt. Eng.*, **28**(2), 90–9 (1989).

T. J. Cloonan, G. W. Richards, F. B. McCormick & A. L. Lentine, 'Extended generalised shuffle network switching architectures for free-space photonic switching', *OSA Proceedings on Photonic Switching*, H. Scott Hinton and Joseph W. Goodman, eds. Optical Society of America, Washington, DC, Vol. 8, pp. 43–7, (1991).

CODE V Introductory User's Guide (ver. 7.2), Optical Research Associates, Pasadena, CA. 1989.

D. L. Crawford, G. W. Taylor & J. G. Simmons, 'Optoelectronic transient response and *n*-channel double heterostructure optoelectronic switch', *Appl. Phys. Lett.*, **52**, 863–5 (1988).

L. A. D'Asaro, L. M. F. Chirovsky, E. J. Laskowski, S. S. Pai, T. K. Woodward, A. L. Lentine, R. E. Leibenguth, M. W. Focht, J. M. Freund, G. Guth & L. E. Smith, 'Batch fabrication and testing of GaAs–AlGaAs field effect transistor self electro-optic effect device (FET-SEED) smart pixel arrays', *IEEE J. Quantum Electron.*, **29**(2), 670–7 (1993).

W. Dobbelaere, D. Huang, M. S. Unlu & H. Morkoc, 'AlGaAs/GaAs multiple quantum well reflection modulators grown on Si substrates', *Appl. Phys. Lett.*, **53**, 9496–9 (1988).

T. J. Drabik & M. A. Handschy, 'Silicon VLSI ferroelectric liquid-crystal technology for micropower optoelectronic computing devices', *Appl. Opt.*, **29**(35), 5220–3 (1990).

F. S. Felber & J. H. Marburger, 'Theory of nonresonant multistable optical devices', *Appl. Phys. Lett.*, **28**, 731 (1976).

M. R. Feldman, C. C. Guest, T. J. Drabik & S. C. Esener, 'Comparison between electrical and free-space optical interconnects for fine grain processor array based on interconnect density capabilities', *Appl. Opt.*, **28**(18), 3820–9 (1989).

W. Franz, 'Influence of an electric field on the optical absorption edge', *Z. Naturforsch.*, **13a**, 484 (1958).

H. M. Gibbs, S. L. McCall, T. N. C. Venkatesan, A. C. Gossard, A. Passner & W. Wiegmann, *Appl. Phys. Lett.*, **35**, 6 (1979).

J. W. Goodman, 'Fan-in and fan-out with optical interconnections', *Optica Acta*, **32**(12), 1489–96 (1985).

M. J. Goodwin, A. J. Moseley, M. Q. Kearley, R. C. Morris, C. J. G. Kirkby, J. Thompson, R. C. Goodfellow & I. Bennion, 'Optoelectronic component arrays for optical interconnection of circuits and subsystems', *J. Lightwave Technology*, **9**(12), 1639–45 (1991).

K. W. Goossen, G. D. Boyd, J. E. Cunningham, W. Y. Jan, D. A. B. Miller, D. S. Chemla & R. M. Lum, 'GaAs–AlGaAs multiquantum well reflection modulators grown on GaAs and silicon substrates', *IEEE Phot. Tech. Lett.*, **1**, 304–6 (1989).

J. Jahns & M. J. Murdocca, 'Crossover networks and their optical implementation', *Appl. Opt.*, **27**(15), 3155–60 (1988).

I. Janossy, J. G. H. Mathew, E. Abraham, M. R. Taghizadeh & S. D. Smith, 'Dynamics of thermally induced optical bistability', *IEEE J. Quant. Elect.*, **QE-22**, 2224 (1986).

F. V. Karpushko & G. V. Sinitsyn, 'The anomalous nonlinearity and optical bistability in thin-film interference structures', *Appl. Phys. B*, **28**, 137 (1982).

L. V. Keldysh, 'Effect of a strong electric field on the optical properties of nm-conducting crystals' *SOV Phys. JETP*, **7**, 788 (1958).

F. Kiamilev, S. C. Esener, V. H. Ozguz & S. H. Lee, 'Programmable optoelectronic multiprocessor system', in *Digital Optical Computing*, R.A. Athale, ed., *SPIE Critical Review Series* CR35 SPIE, Bellingham, WA, pp. 197–200 (1990).

S. H. Lee, S. C. Esener, M. A. Title & T. J. Drabik, 'Two-dimensional silicon/PLZT spatial light modulators: design considerations and technology', *Opt. Eng.*, **25**(2), 250–60 (1986).

A. L. Lentine, H. S. Hinton, D. A. B. Miller, J. E. Henry, J. E. Cunningham & L. M. F. Chirovsky, 'Symmetric self-electro-optic effect device: Optical set-reset latch, differential logic gate, and differential modulator/detector', *IEEE J. Quantum Electron.*, **QE-25**, 1928–36 (1989a).

A. L. Lentine, L. M. F. Chirovsky, L. A. D'Asaro, C. W. Tu & D. A. B. Miller, 'Energy scaling and sub-nanosecond switching of symmetric self electro-optic effect devices', *IEEE Photon. Tech. Lett.*, **6**, 129–31 (1989b).

A. L. Lentine, D. A. B. Miller, J. E. Henry, J. E. Cunningham, L. M. Chirovsky & L. A. D'Asaro, 'Optical logic using electrically connected quantum well p-i-n diode modulators and detectors', *Appl. Opt.*, **29**, 2153–63 (1990).

J. N. Mait, 'Review of Multi-phase Fourier Grating Design for Array Generation', in *Computer and Optically Formed Holographic Optics, Proc. SPIE*, Vol. 1211, SPIE, Bellingham, WA, pp. 67–78 (1990).

F. B. McCormick, 'Free-space interconnection techniques', Chapter 4 in *Photonics in Switching*, J.E. Midwinter ed., Academic Press, Cambridge, MA (1993).

F. B. McCormick & M. E. Prise, 'Optical circuitry for free space interconnections', *Appl. Opt.*, **29**(14), 2013–18 (1990).

F. B. McCormick, F. A. P. Tooley, T. J. Cloonan, J. L. Brubaker, A. L. Lentine, R. L. Morrison, S. J. Hinterlong, M. J. Herron, S. L. Walker & J.M. Sasian, 'S-SEED based photonic switching network demonstration', *OSA Proceedings on Photonic Switching*, H. Scott Hinton & Joseph W. Goodman, eds., Optical Society of America, Washington, DC, Vol. 8, pp. 48–55 (1991a).

F. B. McCormick, F. A. P. Tooley, J. L. Brubaker, J. M. Sasian, T. J. Cloonan, A. L. Lentine, S. J. Hinterlong & M. J. Herron, 'Optomechanics of a free-space switch: the system', *Optomechanics and Dimensional Stability*, eds. R.A. Paquin and D. Vukobratovich *Proc. SPIE*, SPIE, Bellingham, WA, Vol. 1533, pp. 97–114 (1991b).

F. B. McCormick, F. A. P. Tooley, T. J. Cloonan, J. L. Brubaker, A. L. Lentine, R. L. Morrison, S. J. Hinterlong, M. J. Herron, S. L. Walker & J. M. Sasian, 'Experimental investigation of a free-space optical switching network using S-SEEDs', *Appl. Opt., Special Issue on Optical Computing*, August (1992).

D. J. McKnight, D. G. Vass & R. M. Sillito, 'Development of a spatial light modulator: a randomly addressed liquid-crystal-over-NMOS array', *Appl. Opt.*, **28**(22), 4757–62 (1989).

Melles Griot Laser Scan Guide, copyright Melles Griot (1987).

D. A. B. Miller, US Patent No. 4546244 (1985); US Patent No. 4716449 (1987).

D. A. B. Miller, 'Optics for low-energy communication inside digital processors: quantum detectors, sources and modulators as efficient impedance converters', *Optics Letters*, **14**(2), 146–8 (1989).

D. A. B. Miller, D. S. Chemla & S. Schmitt-Rink, 'The relation between electroabsorption in semiconductors and in quantum wells: the quantum confined Franz–Keldysh effect', *Phys. Rev. B*, **33**, 6976–81 (1986).

D. A. B. Miller, D. S. Chemla & S. Schmitt-Rink, 'Electric field dependence of optical properties of semiconductor quantum wells: physics and applications', in *Optical Nonlinearities and Instabilities in Semiconductors*, H. Haug ed. Academic Press, New York, NY (1988).

D. A. B. Miller, D. S. Chemla, T. C. Damen, A. C. Gossard, W. Wiegmann, T. H. Wood & C. A. Burrus, 'Band-edge electroabsorption in quantum well structures: the quantum confined Stark effect', *Phys. Rev. Lett.*, **53**, 2173–5 (1984).

D. A. B. Miller, M. D. Feuer, T. Y. Chang, S. C. Shunk, J. E. Henry, D. J. Burrows & D. S. Chemla, 'Integrated quantum well modulator, field effect transistor and optical detector', *IEEE Photon. Technol. Lett.*, **1**, 62–4 (1989).

K. B. Nichols, B. E. Burke, B. F. Aull, W. D. Goodhue, B. F. Gramstorff, C. D. Hoyt & A. Vera, 'Spatial light modulators using charge-coupled-device addressing and electroabsorption effects in GaAs/AlGaAs multiple quantum wells', *Appl. Phys. Lett.*, **52**(14), 1116–18 (1988).

G. R. Olbright, R. P. Bryan, T. M. Brennan, K. Lear, W. S. Wu, J. L. Jewell & Y. H. Lee, 'Surface-emitting laser logic', *OSA Proceedings on Photonic Switching*, H. Scott Hinton and Joseph W. Goodman, eds., Optical Society of America, Washington, DC, Vol. 8, pp. 247–9 (1991).

U. Olin, 'Model for optical bistability in GaAs/AlGaAs Fabry-Perot etalons including diffraction, carrier diffusion and heat conductor', *JOSA B*, **7**, 35 (1990).

J. W. Parker, 'Optical interconnection for advanced processor systems: a review of the ESPRIT II OLIVES program', *J. Lightwave Technology*, **9**(12), 1764–73 (1991).

J. L. Pezzaniti & R. A. Chipman, 'A polarization metrology for optical interconnects which use polarization beam combining', *Optical Computing, 1991, Technical Digest Series*, Optical Society of America, Washington, DC, Vol. 6, pp. 156–9 (1991).

J. L. Pezzaniti & R. A. Chipman, 'Angular dependence of polarizing beam splitter cubes', *Opt. Eng.*, **35**(5), 1543–9 (1994).

M. E. Prise, N. Streibl & M. M. Downs, 'Optical considerations in the design of a digital computer', *Opt. and Quant. Electron.*, **20**, 49–77 (1988).

M. E. Prise, N. C. Craft, R. E. LaMarche, M. M. Downs, S. J. Walker, L. A. D'Asaro & L. M. Chirovsky, 'Module for optical logic circuits using symmetric self electrooptic effect devices', *Appl. Opt.*, **29**(14), 2164–70 (1990).

D. A. Roberts & R. J. Brown, 'Effect of apertures on Gaussian beams', *Current Developments in Optical Engineering II, Proc. SPIE*, SPIE, Bellingham, WA, Vol. 818, pp. 232–9 (1987).

N. Streibl, 'Beam shaping with optical array generators', *J. Modern Optics*, **36**(12), 1559–73 (1989).

G. J. Swanson, 'Binary optics technology: the theory and design of multi-level diffractive optical elements', *MIT Lincoln Laboratory Tech. Rep.*, **854** (DITC# AD-A-213404) August (1989).

G. J. Swanson, 'Binary optics technology: theoretical limits on the diffraction efficiency of multilevel diffractive optical elements', *MIT Lincoln Laboratory Tech. Rep.*, **914**, (DITC# AD-A-235404) March (1990).

G. W. Taylor, R. S. Mand, J. G. Simmons & A. Y. Cho, 'Ledistor: a three-terminal double heterostructure optoelectronic switch', *Appl. Phys. Lett.*, **50**, 338–40 (1987).

G. W. Taylor, J. G. Simmons, A. Y. Cho & R. S. Mand, 'A new double heterostructure optoelectronic switching device using molecular beam epitaxy', *J. Appl. Phys.*, **59**, 596–600 (1989).

F. A. P. Tooley & A. L. Lentine, 'Digital Optics', Chapter 4 in *Principles of Modern Optical Systems*, Vol. 2, D. Uttamchandani and I. Andonovic, eds. Artech House, Boston, MA (1992).

F. A. P. Tooley & F. B. McCormick, 'Design Issues in free-space Digital Optics', *OSA Proceedings on Photonic Switching*, H. S. Hinton and J. W. Goodman, eds., Optical Society of America, Washington, DC, Vol. 8, pp. 38–42 (1991).

F. A. P. Tooley, T. J. Cloonan & F. B. McCormick, 'On the use of retroreflector array to implement crossover interconnections between arrays of S-SEED logic gates', *Opt. Eng.*, **30**(12), 1969–75 (1991).

A. Vasara, M. R. Taghizadeh, J. Turenen, J. Westerholm, E. Nopponen, H. Ichikawa, J. M. Miller, T. Jaakkola & S. Kuisma, 'Binary surface-relief gratings for array illumination in digital optics', *Appl. Opt.* **31**(17), 3320–36 (1992).

A. C. Walker, 'Reflection bistable etalons with absorbed transmission', *Optics Communications*, **59**, 145–50 (1986a).

A. C. Walker, 'Application of bistable optical logic gate arrays to all-optical digital parallel processing', *Appl. Opt.*, **25**(10), 1578–85 (1986b).

A. C. Walker, 'A comparison of optically nonlinear phenomena in the context of optical information processing', *Optical Computing and Processing*, **1**(1), 91–6 (1991).

S. J. Wein & W. L. Wolfe, 'Gaussian-apodised apertures and small-angle scatter measurement', *Opt. Eng.*, **28**(3), 273–80 (1989).

W. T. Wellford, '*Aberrations of Optical Systems*', p. 116, Adam Hilger Ltd., Bristol (1986).

P. Wheatley, P. J. Bradley, M. Whitehead, G. Parry, J. E. Midwinter, P. Mistry, M. A. Plate & J. S. Roberts, 'Novel nonresonant optoelectronic logic device', *Electron. Lett.*, **23**, 92 (1987).

B. S. Wherrett, 'Semiconductor optical bistability: toward the optical computer', in *Nonlinear Optics: Materials and Devices*, C. Flytzanis & J.L. Oudar, eds., Springer-Verlag, Berlin (1986).

B. S. Wherrett, A. C. Walker & F. A. P. Tooley, 'Nonlinear refraction for cw optical bistability' in *Optical Nonlinearities and Instabilities in Semiconductors*, H. Haug ed. Academic Press, New York, NY (1988).

T. K. Woodward, L. M. F. Chirovsky, A. L. Lentine, L. A. D'Asaro, E. J. Laskowski, M. Focht, G. Guth, S. S. Pei, F. Ren, G. J. Przybylek, L. E. Smith, R. E. Leibenguth, M. T. Asom, R. F. Kopf, J. M. Kuo & M. D. Feuer, 'Operation of a fully integrated GaAs–AlGaAs FET-SEED: A basic optically addressed integrated circuit', *Phot. Tech. Letts.*, **4**, 614 (1992).

E. Yablonovich, T. J. Gmitter, J. P. Harbison & R. Bhat, 'Extreme selectivity in the liftoff of epitaxial films', *Appl. Phys. Lett.*, **51**(26), 2222–4 (1987).

E. Yablonovich, T. J. Gmitter, J. P Harbison & R. Bhat, 'Double heterostructure GaAs/AlGaAs thin film diode lasers on glass substrates, *IEEE Phot. Tech. Lett.*, **1**, 41–2 (1989).

Index

271